P9-AQD-952

WHY DOES MY CAT . . . ?

WHY DOES MY CAT . . . ?

Sarah Heath

Foreword by John Fisher
Author of *Why Does My Dog . . . ?*

Drawings by Russell Jones

SOUVENIR PRESS

To all my family
for their unending love, support and encouragement

First published 1993 by Souvenir Press Ltd,
43 Great Russell Street, London WC1B 3PA

Reprinted 1997
Reprinted 1998

ISBN 0 285 63147 0

Photoset by Rowland Phototypesetting Ltd,
Bury St Edmunds, Suffolk

Printed and bound in Great Britain by
Biddles Ltd, Guildford and King's Lynn

CONTENTS

FOREWORD

A recent report stated that dog ownership in Britain has stayed level at 7.3 million, whilst cat ownership has risen to seven million. This confirms the predictions made some years ago, that cats will soon exceed dogs as the number one companion animal. In this extremely readable book Sarah Heath manages to pack in so much knowledge that very few of the seven million (and rising) cat owners will fail to understand why their cat behaves as it does. But this book does more than improve owner understanding: it increases our admiration for this affectionate, yet often aloof animal who chooses to share its life with us mere humans.

Whilst reading *Why Does My Cat . . . ?* three unwritten facts became crystal clear to me about Sarah Heath. The first was just how deep her understanding of feline behaviour really is; the second was how much her veterinary background has helped her to analyse cats' behaviour and describe practical treatment programmes; and the third was how her deep respect and obvious love for cats allows her to answer the owners' questions with a reply that carries the message *I understand how you feel and I sympathise with you.*

Within the pages of this book, almost every recognisable behaviour exhibited by our friend the cat is described, and some of the solutions advised by Sarah for overcoming problem behaviour are surprisingly simple. I am sure that many an owner has slapped her forehead and said, *'Of course! Why didn't I think of that?'*—like the case of Trinder who had started house soiling, until Sarah noticed that her litter tray had been placed too close to the bed of a nine-stone Bull Mastiff. Obvious, when we are told, but we often don't see the wood for the trees. Some cases are much more difficult and I couldn't help wondering, whilst reading about Jake (the local terrorist) who breaks into houses through the cat flaps, beats up the resident feline and leaves his calling card by spraying all over the house, just how many of his victims Sarah would be treating at the same time.

As well as being a very comprehensive book on feline behaviour problems (the cause, effect and cure), we are also treated to a fascinating history, not only of cats, but also of man's changing attitude towards them. By describing how, over the years, we have tortured them, burnt them to death in their thousands, interred them in the walls of houses and ritually thrown them off 230-foot-high towers, and how many of us today have come to rely upon them for affection and companionship, Sarah effectively exposes a chink in our own personality profiles. Throughout these changing attitudes the cat has continued to try and domesticate *us*—a fact that comes over clearly in the various chapters: cats live with humans because it suits them to do so.

Why Does My Cat . . . ? is a wonderful book that will rank high amongst the modern books on companion animal behaviour. As the author of *Why Does My Dog . . . ?* I look forward to seeing both books side by side on the shelves, and I shall feel honoured to have my book placed alongside Sarah Heath's.

John Fisher
Author of
Why Does My Dog . . . ?

ACKNOWLEDGEMENTS

Many thanks to my husband Philip and son David for their patience and understanding during the writing of this book, and to my wonderful parents who have always been there to give me support and encouragement. Thanks also to all my colleagues in the Association of Pet Behaviour Counsellors (APBC) for guiding me in the right direction. Special thanks to John Fisher for his support and constant good humour, to David Appleby for introducing me to the Association and to Claire Bessant for her advice and reassuring motivation. I would like to acknowledge Peter Neville's enormous contribution to the understanding of feline behaviour and express my sincere gratitude to him for encouraging me to write this book. A special mention goes to Matthew James who has been a great comfort when the going has been tough and has inspired me to keep going. Finally I must thank all the cats that I have worked with, and especially my own two, Truffles and Gremlin, for everything that they have taught me about cat behaviour.

Part One

BEHAVIOUR THERAPY FOR CATS

1

WHAT IS A BEHAVIOURIST?

When people mention the subject of pet behaviour and combine that with the use of the word 'therapist', it is almost guaranteed to induce stifled giggles and knowing glances from those around. For most people behaviour therapy invokes images of a pet lying on a comfortable Dralon couch while a bespectacled eccentric asks it deep and meaningful questions about its disturbed 'childhood'. The worst extremes of the caricatured psychiatrist are combined with the anthropomorphic cartoon strip, to arrive at a pet behaviourist image that arouses uncontrollable laughter.

Thankfully, the reality is very different. Pet behaviour therapy is an important and ever-increasing part of the overall support service for pet owners. The causes of feline behaviour problems are numerous and in my capacity as a behaviour counsellor I offer time and patience to listen to owners' concerns regarding their cats. Then, using my knowledge of feline behaviour, I endeavour to establish the root cause of the problem and design a behaviour modification programme that is suited to the individual needs of the cat and its owner. The ultimate aim of any pet behaviour therapist is the establishment of a mutually beneficial relationship between owners and their pets. In many cases the lives of pet and owner are inextricably linked, and the strains and stresses of the family have a profound effect on the behaviour of the cat. It is therefore essential that behaviourists have a patient and compassionate approach in dealing with what is often a very sensitive matter.

A behaviour problem is often only recognised as such when the strain of dealing with it starts to overshadow the pleasure of owning the cat. Many distressed cat owners find it difficult to know where to turn for help, and often friends and relatives who also own cats are the first port of call. Abnormal behaviour is often seen as an indictment of the owner's ability to care for the animal. It produces feelings of inadequacy, guilt and failure in the owners who blame themselves for the breakdown in their relationship with their pets. Just as society finds it difficult to talk openly about mental and behaviour problems in people, so pet owners feel unable to admit openly that all is not well with their beloved cat.

If friends and relatives are unable to offer any consolation or advice, the veterinary surgeon is often the next stop. On many occasions I have answered the telephone to be greeted with the words 'you are my last hope'. Changes in behaviour are frequently the reason for cats being brought to the surgery and of course some abnormal behaviour patterns have medical causes. Such problems will be detected by a full medical examination, but where there is no evidence of a medical problem it will invariably take a great deal of time and patience to consider all the relevant information and establish a cause. Some veterinary surgeons, like myself, have a particular interest in animal behaviour and can set aside the required time for a full investigation of behaviour problems, but for many vets the pressures of a busy general practice make it impossible to devote sufficient time to these cases. For this reason it is essential that pet behaviourists specialising in this field should work alongside veterinary surgeons and provide a referral service. In this way, after carefully excluding any medical problems, the veterinary surgeon can refer his client to someone who has the time and experience to suggest practical ways of resolving the problem.

The Association of Pet Behaviour Counsellors (APBC) provides just such a service, with a nationwide network of behaviourists who work solely on referral from veterinary surgeons. Since its formation in 1989, the Association has grown considerably and now consists of 15 United Kingdom members with some 35 to 40 clinics being held throughout Britain and Ireland. Many veterinary practices provide premises for these clinics, and indeed three of the university veterinary schools incorporate clinics run by APBC members. As the demand for these services increases it is hoped that the APBC will be regarded as the recognised body of pet behaviourists and will form part of a readily available nationwide referral service for pet owners.

You only need to watch your cat for a short time to realise

that the behavioural repertoire of this wonderful creature is immense and varied, so it comes as no surprise to be told that the range of behaviour problems encountered by the APBC is also vast. One of the great privileges and benefits of working alongside the animal kingdom is that there is never a shortage of new experiences, and by pooling the knowledge of the behaviourists within the APBC we can widen our reference material and be better equipped to deal with the bizarre and the unusual. When a Red Burmese called Murphy was brought into our surgery because he periodically attempted to tear out his own tongue, it certainly got everyone thinking! Although the effects of the behaviour could be halted by the use of mittens applied to all four feet, the fascinating question was, why should a cat deliberately mutilate itself in this way? These episodes of self-inflicted trauma were dramatic and during an attack the self-mutilation was intense, but despite the obvious pain that the cat was causing itself it continued to claw furiously.

Obsessive compulsive disorders (OCD) such as these are recognised in humans and are also found in the domesticated species where, according to some sources, cats are more heavily represented than dogs. In the treatment of such cases behaviourists follow the developments in the human field where the environmental and managemental stress factors, which are contributing to the condition, are removed as far as possible. Medical treatment is also available and its effectiveness seems to be influenced by the presence of conflicts as well as by the time during which the OCD has been apparent. As yet knowledge about the causes and successful treatment of these cases is limited, and so the pooling of expertise within an organisation such as the APBC is invaluable.

The APBC has five overseas member practices—in Denmark, Belgium, Switzerland, Canada and the United States. The combination of these and the practices throughout the United Kingdom provides extensive sources of information. Dr Don McKeown, who works at the University of Guelph in Ontario, Canada, has done much work on OCD in the domestic species and was therefore able to offer expert advice on Murphy's apparently masochistic problem. (More about Murphy in Part Three.)

Animal behaviour therapy is a relatively new science but a rapidly expanding one. It is essential that the wealth of knowledge about behaviour problems should be made available to pet owners, and the APBC is doing just that. I hope that in the following pages you will discover more about the way your cat behaves and so increase your understanding of this remarkable creature.

2

WHAT AFFECTS BEHAVIOUR?

DIET

For many years in the veterinary world the cat was regarded as a smaller version of the dog and it is only relatively recently that feline medicine has become an important discipline in its own right. We now have a wealth of information available in this field and we know that the cat is indeed a fascinating and unique species.

One area in which the difference between the cat and the dog is most noticeable is that of nutrition. The provision by pet food manufacturers of a wide range of foods specifically for cats, in addition to those intended for dogs, is not merely a marketing ploy but rather the result of the very different nutritional requirements of the two species. However, a surprising number of pet owners feed their cats on dog food, due to the fact that dog food is cheaper, unaware of the potential harm they may be doing to their cats. Cats and dogs are both carnivores but the cat does not possess some of the enzyme pathways that are found in the dog and it is therefore incapable of synthesising some nutrients in its body from the components in its diet. Examples include vitamin A and a substance called taurine, both of which do not occur in vegetable matter. It is therefore essential for at least part of the cat's diet to be of animal origin. A truly vegetarian diet is not suitable for the cat, and preparatory dog foods which do not have sufficient levels of essential feline nutrients are equally inappropriate.

In extreme cases, when cats have been fed on unsuitable diets for some time, their health may be affected, the best known example of this probably being the effect on the eye of long-term taurine deficiency. Taurine is necessary for the maintenance of the retina and without it the retina will begin to degenerate. Eventually a cat that is fed on a taurine-deficient diet will lose its sight completely, but it can take quite some time for the owner to realise that there is anything wrong since a cat that is kept indoors can adapt quite well to deficiency in its sight and so conceal the problem. Sometimes the first thing that the owner notices is that the cat is 'behaving strangely', and she may interpret this as a behaviour problem.

One such case involved a tabby cat called Millie who lived in a city in a high-rise block of flats. For reasons of safety she was not allowed access to the outside world and as a result she never had the opportunity to hunt and supplement her diet with animal protein. The owner reported that Millie had begun to act very strangely, over-reacting to sounds in the flat, and at times appearing to be very distant—in fact the family was convinced that she was hallucinating. If we had tried to treat this cat as a behavioural case, without taking the time to get a full history or give the animal a full medical examination, it would have been possible to miss the real cause of the problem altogether.

When I questioned the owners carefully, it emerged that Millie was fed exclusively on a brand of dry dog food. She was never given milk, which does contain a small amount of taurine, since they had been told when she was a kitten that the milk that they were giving her was causing her persistent diarrhoea. All in all Millie's diet was totally unsuitable, and a full medical examination revealed changes in the retina that pointed to taurine deficiency. Cases like this underline the need for veterinary involvement in all 'behavioural' cases and explain why all members of the Association of Pet Behaviour Counsellors see cases only on referral from veterinary surgeons, so that any medical reasons for their behaviour can be identified first.

Links between 'behaviour' problems and diet may not always be so obvious and it is true that cats fed on dog foods are not the only ones who present problems connected to the diet. Not all preparatory cat foods seem to be suitable for all cats and in some cases questioning the owner can reveal a strong link between the behaviour and the diet. Problems that always occur at a specific time interval after feeding, and those that have the hallmarks of an allergic reaction, will ring

Links between behaviour problems and diet may not always be obvious.

alarm bells for the behaviourist and a change in diet may be suggested.

One example of this involved a rather quiet and withdrawn individual called Tabitha. According to her owner, Tabitha had been a very loving and affectionate companion, but gradually, over the previous few months, she had become less and less so. The situation had now reached the stage where Tabitha went through what the owner called a variety of mood swings throughout the day. First thing in the morning she would be reasonably affectionate and would ask for food as she had always done. As the morning went on she would become more and more offhand and irritable, and if the owner tried to cuddle her she would even become aggressive. By lunchtime her behaviour had deteriorated to its lowest level, and the owner found that the only thing to do was to ignore her. During the afternoon Tabitha would begin to mellow and by early evening was once again her amiable self. She would then start to brush up to her owner and ask for food in her inimitable way. However, as the evening progressed the pattern of behaviour experienced during the day would be repeated and, as you can imagine, the relationship between Tabitha and her owner was becoming strained.

When I questioned the owner, there did not seem to have been any relevant changes in Tabitha's life that would explain the alteration in her behaviour. However she did mention that there had been changes in Tabitha's coat and a noticeable increase in hair loss, and that these had started at about the same time as the behaviour changes. Such coat changes may be an indication of an allergic reaction and the fact that the behaviour pattern seemed to show a deterioration after feeding and a slight improvement before was added endorsement. Despite the fact that Tabitha had always been fed on the same brand of tinned cat food, I suggested a short-term diet change. Within just a few days the owner telephoned to report a marked improvement in Tabitha's behaviour and a few days later she reported a dramatic decrease in hair loss and an improvement in the cat's coat.

Although there is, as yet, no hard scientific evidence for a link between behaviour problems and diet, circumstantial evidence in support of this theory, such as Tabitha's case, is growing. We do not know why such reactions occur. It could be that Tabitha's age and changes in metabolism had resulted in an allergic reaction to her usual diet, or that the manufacturers had altered some ingredient. It is even possible that Tabitha's immune system had become sensitised over the years to some allergen in her diet. Whatever the reason, her owner was relieved to have her old cat back and the actual mechanics involved seemed somewhat irrelevant. Certainly, the links between diet and behaviour are an area in which further research is needed, and I hope that in the next few years more information will become available.

ENVIRONMENTAL INFLUENCES

There can be no doubt that human lifestyles have altered quite dramatically over the years. Despite the recent period of recession, some might even say depression, we in the West do still enjoy a relatively affluent way of life. This increase in affluence has brought with it a dramatic move away from the quiet rural and agricultural lifestyle to the hard, ambitious life of city businesses. Population density is rapidly increasing and the world is becoming an ever more impersonal and lonely place to be.

Along with the changes in our lifestyles, we have seen a noticeable alteration in our motives for keeping pets, which have contributed to a dramatic increase in the popularity of the cat. We know from research that pet ownership is good for us in medical terms, and that decreases in both heart rate and blood pressure are attributed to stroking our pets. We also recognise that there are social benefits, and unfortunately over recent years there has even been a tendency for some of the canine breeds to be regarded as some kind of status symbol. The dog has traditionally been viewed as the number one companion and it has earned this position by its ability to assimilate into the human social group. In an ideal situation the dog regards his human packmates as dominant leaders and happily accepts a subordinate role while still being prepared to share in the protection of the den. The protective aspects of canine behaviour are probably the most common reason for people deciding to keep a dog and most of us encourage this behaviour in our dogs, at least to some extent. In our increasingly hostile society, an added feeling of security in our homes or out on walks is welcome.

However, in the more affluent sections of society the dog has been superseded by a barking electronic security device, and elaborate alarm systems are installed to protect property.

The dog has been manipulated by man to fulfil numerous other tasks, and the enormous variety in size, shape and temperament are evidence of this. Training of dogs to aid in the detection of drugs and bombs and to assist the blind, disabled and elderly members of our society helps to increase society's acceptance of canines in the community.

As the recession bites, however, fewer people can afford to keep a dog purely as a pet and those that can usually find that their hectic social calendar leaves little time for canine companionship. People want the affection of a pet without the ties and the effort that dogs demand, and the cat fits much more easily into the pattern of late nights at the office and then straight on to the night club. The poor old dog would be crossing his legs, but the independent cat sees to his own needs and the owner can return home free from guilt and enjoy the company of his feline friend without the demands. Equally, as the trend in housing in this country moves away from large, old family houses to compact flats and starter homes, the dog is not so readily accommodated while the cat who, in the words of the poem, 'sleeps anywhere', can easily share our den.

Society conditions us to accept that self-sufficiency is of

paramount importance, and that displays of emotion indicate weakness and a lack of resolve. However, the truth is that we all need an outlet for our caring and nurturing qualities, and as our children mature earlier and earlier we turn more and more to our pets for unconditional displays of affection. For the young and relatively wealthy members of our society in particular, the cat is the ideal pet for their unpredictable lifestyle; for the cat, on the other hand, this assimilation into modern life is not always straightforward.

Young couples in exclusive London flats tend to want to compartmentalise their cat in the same way as they do themselves. When you enter their homes you may find the cat's belongings—a bed, a food bowl and of course a litter tray—all neatly tucked away in one corner of the tiny kitchen. That may be all very well for the owners and may add to the ultra-organised and efficient image that they wish to convey to the outside world, but from the cat's point of view the arrangement is often unworkable. Few of us would be prepared to eat our dinner in the lavatory, so why should we expect our cats to dine in close proximity to their latrine? For a species that is renowned for its cleanliness, it seems logical that the cat will separate feeding and fouling in the only possible way and move elsewhere to urinate and defaecate, so creating a totally preventable problem. It may sound obvious when put like that, but for the high-flying business executive who returns to find cat mess on his priceless Persian rug, logical explanations often provide little comfort!

Modern urban life and increasing affluence not only result in changes of attitude to how the cat should live within the home, but also to how much interaction the cat should have with the outside world. It would appear that a larger percentage of feline behaviour cases are being referred from members of the pedigree section of the cat population than from the dear old mog. This is partly a result of the fact that these cats often cost their owners a considerable amount of money and therefore the owners are not so likely just to reject the cat and get a replacement, which unfortunately happens regularly with mogs. It is also a result of the changing attitudes to cat-keeping and the increasing tendency to house cats, especially pedigree ones, permanently indoors. In areas where traffic poses a serious threat, many owners do not feel they can risk allowing their valuable pet free access to outdoors. Consequently more and more cats are becoming exclusively house pets and are facing the prospect of a life without the great outdoors, deprived of the opportunity to put those remarkable hunting instincts into practice. By no means all cats that are kept permanently indoors suffer from related behaviour problems: the majority who enter this existence as kittens and never experience the joys of the outdoor world, do adapt very well to indoor life. We may argue that the owners are missing out by never witnessing the true elegance of a cat on the prowl or the murderous precision of a hunting feline, but being denied the opportunity to express these natural behaviours does not seem to bother the cats and their sedentary and insulated lifestyle is perfectly acceptable to them.

For others, however, the strains of suppressing the wild side of their nature can become too much, and problem behaviours can result from a lack of outlets for their inherited natural feline behaviour. Cats in general and some breeds, such as the Burmese, in particular, are very intelligent and consequently demanding creatures.

One such cat was Rumble, a three-year-old Burmese who lived in a select apartment. He was dearly loved by his owners, David and Chloe, and led a totally indoor life for fear of the main road which was only a matter of yards from the apartment block. Rumble was on the whole a loving individual who repayed his owners' kindness with frequent displays of affection, but occasionally, when they returned home after both working late at the office, he would turn on them with even more convincing displays of aggression. For David and Chloe the contrast in Rumble's behaviour was extremely distressing and they found that these occasional outbursts were marring the whole relationship with their cat. In Rumble's case the possibility of a link to diet was ruled out and we took a closer look at his lifestyle for clues to the cause of his problem. Despite the fact that he was undoubtedly much loved and certainly well fed and cared for, his life in the apartment provided little stimulation. For such an intelligent and active species, this lack of challenge resulted in a very low excitement threshold, and the contrast between solitude in a quiet apartment during the day and the company of his doting owners in the evening was enough to induce a staggering over-reaction which manifested itself as aggression. Treatment obviously needed to involve providing increased stimulation for Rumble

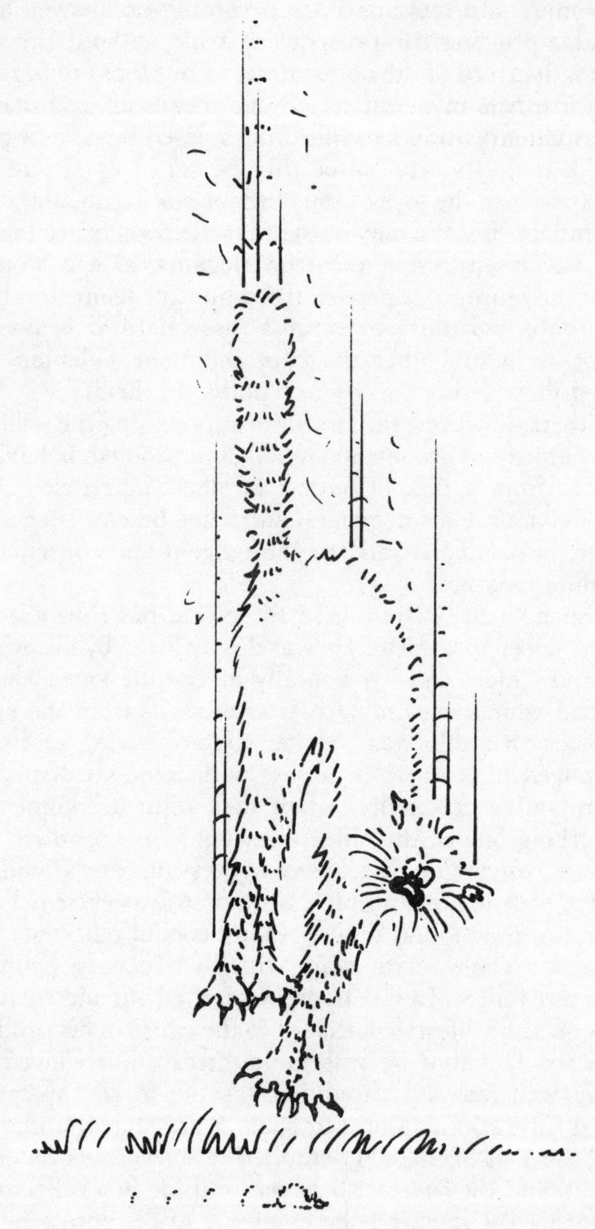

The murderous precision of a hunting feline.

in his owners' absence and once this was done he did make remarkable progress. However, with some cats the apartment lifestyle is just not for them, and there is no alternative but to suggest to the owners that it would be kinder to rehome these individuals away from the city lights.

3

THE HISTORY OF OUR MODERN CAT

During its long history of association with man the cat's reputation has swung wildly from one extreme to the other. In ancient Egypt it was regarded as a divine creature worthy of reverence and worship, but subsequently, in medieval Europe, it became an object of utter revulsion due to its links with witchcraft and the devil. Nowadays cats are highly popular domestic pets throughout the world, but they still invoke extreme feelings of both love and hatred and few people feel indifferent towards them.

The dictionary definition of 'domestication' is 'to tame or bring under control', and most cats would object most strongly to the implication that they are dependent on man for their survival. They regard themselves as free agents, and whilst accepting the food and shelter that we offer they retain their independence and, in most cases, still display their proficient hunting skills whenever the opportunity arises. It is true that man has manipulated the cat to some extent and has bred for tolerance, affection and tameness. However, the mere fact that so many cats have unlimited access to the great outdoors makes it impossible for humans to be totally in control of their breeding programmes and the cat has managed to escape relatively unscathed from selective breeding. We see very little variation in size or structure, and although the coat may vary substantially in colour and length, we do not see the extreme breed variations that are displayed in the canine world.

We do not know exactly where, when or indeed how the process of feline domestication began, but early records of an association between man and cat in Ancient Egypt date from around 2600 BC; although cat remains have been found in other locations there does not appear to be any firm evidence of domestication outside Egypt at that time. Early pictorial references to the cat can be found in Egyptian tomb paintings dating from as long ago as 2000 BC, when the artists' subject appears to have been the jungle cat (*Felis chaus*). The famous painting of a cat in a papyrus swamp, from the tomb of Nebamun, dates from the Eighteenth Dynasty (c.1400 BC), and there is general agreement that it may have been a wild cat. References to the cat in the earliest obviously domestic setting can be found in paintings from the tomb of May, dating back to c.1450 BC, in which the cat is seen tethered to a chair leg. Later works, such as wall paintings from tombs at Deir el Medina dating from c.1275 BC, show the cat in family settings.

All this pictorial evidence would seem to point to the beginning of the Eighteenth Dynasty as the time of feline domestication. However, such conclusions lead to debate amongst researchers, since this was also the time when the cat became established as a religious symbol in Ancient Egypt and many argue that the human-feline relationship must have been a long-standing one by this time in order for it to achieve such a status.

There is also much mystery and debate surrounding the question of a direct ancestor for our modern cat. The most likely candidate among the twenty-six species of small wild cat throughout the world is the African wild cat (*Felis lybica*). At one time *Felis lybica* was believed to be a species distinct from the European or Scottish wild cat (*Felis silvestris*), but despite their quite different appearances it is now thought that they are opposite extremes of a single species. The southern African wild cat is a relatively fine-limbed specimen, with a small head and large ears, whereas the northern Scottish wild cat is of a heavier build with shorter, stocky limbs and short ears. Both cats have a striped tabby coat, but this is more heavily marked in the northern variety while having a more flecked appearance in the African wild cat.

Evidence for a link between the modern domestic cat and the African wild cat has come from studies both of anatomy and of similarities in coat colour and pattern. It has also been shown that the domestic cat and the relevant sub-species of the wild cat have a common chromosome complement of nineteen pairs, as well as a closely similar brain size as measured by the cranial capacity of the skull. It is well known that domestication in any species results in diminution of the skeleton, and evidence from research carried out by Hemmer in West Germany shows that the brain also decreases in size during the domestication process. The direct ancestor of any domestic species will also have a smaller brain than the other members of the wild species, and in the case of the cat it is the southern race, *Felis lybica*, which has the relatively smaller cranial capacity.

Behavioural links between the domestic cat and the African wild cat have also been cited as evidence that it is *Felis lybica* and not its European cousin who is the true ancestor of our fireside companions. Work carried out with the Scottish wild cat has shown that these cats are virtually untameable: despite intensive early contact with wild cat kittens the naturalist Frances Pitt found that the most she could achieve was a state of uneasy truce; she could make no real progress towards forming a relationship with them. When she achieved successful cross-matings between one particular wild cat and her domestic cats, she described the offspring as 'nervous and queer tempered' and admitted that they could not be trusted once they had become half-grown.

In contrast, the African wild cat is said to have a greater affinity with people, and in the southern part of Sudan it can be found living in close proximity to the local Azande tribe. There has been no attempt to domesticate the cat in this area, but the mere fact that it is prepared to live close to human settlements provides a sharp contrast with the elusive and unsociable behaviour of its European cousin.

Although evidence points to the African wild cat as the direct ancestor of the domestic cat, recent work has called into question the involvement of the jungle cat (*Felis chaus*). Certainly the Ancient Egyptians appear to have mummified the jungle cat along with the African wild cat, and supporters of its contribution to the domestication process claim that cross-matings between the jungle cat and the domestic cat do result in more human-tolerant offspring, which may have been a useful factor in the early stages of domestication.

Too many questions still remain for any firm conclusions to be drawn; all that seems certain is that the precise mechanics, and indeed the timing, of feline domestication are still a mystery!

It has been suggested that the association between man and cat resulted from a need to protect the large grain stores of Ancient Egypt from rodents. In return for tolerating links with mankind the cats received large quantities of natural prey in the form of rats and mice, and so a mutually beneficial relationship developed. African wild cats that were sufficiently tolerant of people certainly seem to have hunted prey such as rodents and birds in and around the towns, and more confident individuals probably scavenged waste food from the kitchen yards at the back of the town buildings. Thus the town's cat population would steadily grow, supported by the domestic waste of human inhabitants, much like the feral cat populations of today.

How, why or when the cat found its way into early Egyptian life may be uncertain, but we know that it did and we also know that for more than 1,300 years it was regarded by that society as a national deity. It was the feline fertility goddess Bastet who inspired this devotion and the Temple of Bastet at Bubastis in the Nile delta was the focus of cat worship. The goddess's annual festival attracted hundreds of thousands of devotees who flocked to Bubastis from all over Egypt.

Bastet was originally portrayed as a lioness and it is believed that the advent of the domestic cat marks the time when she started to be portrayed in bronze statues as a cat or a cat-headed woman. These statues carried a sistrum, a wire-rattle used in Isis worship, which was believed to sound out the rhythm of love-making, an aegis or shield, and a basket. There were four kittens at her feet and four moving bars in the sistrum, both of which refer to the fact that female cats usually give birth to four offspring. It is likely that cats were kept at temples in Ancient Egypt as incarnations of the fertility goddess and some believe that the first domestic cats resulted not from opportunist town cats but rather from hybrids of these temple cats. The reverence afforded to the feline population was not limited to the temple residents: household cats demanded exactly the same level of respect, and when they died their bodies were taken to Bubastis where they were embalmed and buried in sacred vaults, while the owners would go into mourning and shave off their own eyebrows.

Elaborate linen wrappings for feline mummies appear to have been restricted to cats owned by the rich and famous, but coffins have been found for cats from all sections of Egyptian society. No doubt we could learn much about the history of the cat by studying these mummies, but unfortunately in 1889, when a consignment of 19.5 tons of cat mummies docked at Liverpool, their historical interest was completely overlooked and they were sold off as fertiliser at the equivalent of around £4 per ton! From that consignment only one cat skull remains, now held at the British Museum.

As a result of the movements of both merchants and military units, the domestic cat slowly spread from Egypt and found its way across the world and into Europe. At this time the cat still retained a divine status and it continued to be accompanied by the sistrum which probably formed the basis of portrayals of cats playing the fiddle. Indeed, it would appear that it is this connection that is referred to in the nursery rhyme 'Hey, diddle diddle'.

The links between the cat and the fertility goddess were still evident when the cat reached the shores of Britain around the time of the Roman invasion, and it was not until Christianity was established as the Empire's accepted religion that its status as a divinity was lost. Even though it no longer commanded divine respect, the cat of the medieval period was believed to retain certain magical properties and many stories from this period refer to cats that brought their owners good fortune. This belief was the reason behind the practice that survived right up to the end of the nineteenth century, whereby cats were interred in the walls of houses in order to protect them from evil influences.

Despite these positive interpretations of cat magic, there was also a sinister side which associated the cat with demons and witchcraft. In the late Middle Ages, during a period of unparalleled cruelty to cats, they were burnt to death in their thousands. The cat became a symbol of the devil himself and by burning the creatures alive it was believed that the devil was made to suffer. The cat's reflective eyes which, in Ancient Egypt, were believed to hold the rays of the sun and be a symbol of the feline deity, now became a symbol of demonic powers. As the wave of hysterical persecution against witches swept across Britain, so followed the campaign against their feline companions.

One of the reasons for this misunderstanding came from observations of the cat during the mating process, when the female screams and strikes out at the male as he withdraws. Observers believed that the tom cat's semen must be burning the queen and they saw this as a link with the fire imagery of Satan. What they were in fact seeing was a biological mechanism to induce ovulation in the queen, but this was not recognised until very much later. Unfortunately for the cat, the blinkered outlook of the period saw its sexual habits of multiple matings as a further sign of ungodliness and so added to the hysteria that already existed.

Eventually cats were regarded as the source of witchcraft power. They became labelled as the Witch's Familiar and any ordinary women who owned cats were believed to be using them to carry out evil deeds. The mere fact that a woman owned a kitten could be evidence enough to convict her of witchcraft, and any signs that she had showed kindness to cats would immediately put her under suspicion in the eyes of the courts. As the reputation of the cat sank, so its torture and persecution became accepted as part of everyday life, and even up to the eighteenth century vicious teasing and abuse of cats was widespread.

Gradually, during the mid-eighteenth century, the tide began to turn and the cat started to take on an air of respectability, but misunderstanding continued for some considerable time and this was reflected in the large number of feral cats as opposed to household pets. The very aspects of feline behaviour that make it such a desirable pet today fuelled a reputation for unreliability and untrustworthiness. It was to take many years for the witchcraft hysteria to disappear completely, and the independent nature of the cat, with its aloof expressions, made people reluctant to welcome stray cats into their homes. Due to their inability fully to understand and control the cat in the same way as the dog, they maintained a healthy distance: although the cat was no longer persecuted, it lived alongside humans rather than with them.

It was not until as recently as 1871, when the first British cat show was organised by Harrison Weir, that public opinion finally turned and the cat became established as a favoured domestic pet. Since that time its popularity has steadily increased, and today it is set to become the most numerically popular pet in our land. It is now regarded as an animal to be admired rather than despised, and although we do not consider it worthy of worship, we do look upon it with a certain degree of respect for its grace, agility and independence.

4
ESTABLISHING CAUSES

In the previous chapter we saw how the human–cat relationship has evolved and looked at the maternal basis for our links with the domestic cat of today. We are increasingly recognising the great value that our feline friends have to offer with their ability to combine two distinct facets of their personality. It fascinates us to see how the perpetual kitten and independent adult can exist simultaneously, and while enjoying the moments of affection that our cat offers us, we respect him for the wilder side of his nature. We do not presume to take our cats for granted in the same way as we do our dogs, since deep down we feel privileged to share our lives with them.

It is this difference in attitude that is reflected in the cat owner's reaction to behaviour problems. When a cat fulfils its role as loving pet for ninety per cent of the time, owners may convince themselves that they are in some way to blame for the temporary lapses in behaviour that occur. They feel that the cat must be unhappy in their company and believe that they are inadequate owners. Rather than let anyone think badly of the cat that they adore, they will go to incredible lengths to conceal their problems, and often the deterioration in household hygiene due to problems of house-soiling and indoor marking results in a reluctance to have visitors in their homes.

In the end the owner has less and less contact with other people and his or her relationship with the cat consequently takes on more and more importance. For owners like this rehoming the cat is never an option: despite what are often quite severe problems they persevere in the hope that things will get better. Eventually there may come a time when the strain of coping with the problem becomes too great and not only is the relationship with the cat pushed to breaking point, but relationships within the family also start to show signs of strain.

The level of toleration varies considerably between families. Only last week I received phone calls from two clients with essentially the same problem: both had a house full of cats and both had been experiencing a problem of house-soiling. However, in the one case I was being contacted after the cats had made mistakes on just two occasions, while the other owners had lived with the problem for nearly two years and had only now decided that they needed some help.

Once the caring cat owner has accepted that there is a problem, the next obstacle is whom to approach for help. For most the first port of call will be a friend or relation with a common love of cats, in the hope that they may have experience of a similar problem and be able to offer some words of wisdom. In many cases such advice is all that is needed and the problem is resolved, but owners need to remember that what constitutes a problem depends entirely on the individual: what one considers a serious problem may be viewed by others as perfectly acceptable behaviour. It can take a lot of courage for people to admit that all is not well in their relationship with their pet and it will not help their situation to be told that they are worrying unnecessarily. We need to recognise that once the owner has identified some behavioural trait as a problem, it needs to be taken seriously.

The understanding of behaviour in any species is a complex subject and one that requires a complete picture of the individual concerned. It is important to examine all aspects of the animal's life. In this way we hope to pick up all relevant facts and so piece together a cause of the problem. Failure to take an overall view of the situation can result in jumping to conclusions and failing to identify some vital clue. However, this holistic approach can prove very difficult in the case of the cat where we are dealing with an individual who leads two very separate lives. On the one hand she fulfils the role of loving pet and on the other she is a cruel, calculating killer and a truly wild animal. Unlike the dog owner, who can tell you almost everything that his pet gets up to throughout the day, cat owners are often completely unaware of where their pet goes all day, let alone what she does! Consequently our questions often lead to a blank and we are left trying to piece together a rather fragmented story.

Understanding behaviour in any species is a complex subject.

Obtaining an accurate history is really the key to success in dealing with behaviour problems, and in order to achieve this it is necessary to take the time to carry out detailed consultations. Ideally all the members of the family should be present, since each may have a slightly different perception of the problem and each viewpoint can be vital in completing the whole picture. It can also help to establish any possible links between family tensions and the cat's problem, for our feline friends are extremely sensitive to their environment.

I use the same questionnaire at the start of all consultations and tell the owners that although some of the questions I ask may seem irrelevant to their own concerns, it is important to follow such a set procedure in order to guard against rambling conversations that completely miss the point. The following examples will give you some idea of the detective-like nature of a behaviourist's work:

Where did you acquire the kitten?
The early environment of the kitten has a marked effect on its subsequent behavioural development. The personality of a cat reared in a semi-feral state on the farm will vary considerably from that of a pedigree cat reared in a large breeding cattery. The early experiences that are available are obviously more limited in the latter, and farm-reared cats are often more

confident in themselves. They have had more opportunity to learn through inquisitive exploration and have been free to venture farther from the nest, so increasing their independence. Consequently as adults these cats often earn a reputation for being self-sufficient and aloof.

We know that there is a strong maternal influence on behaviour—toilet training and hunting skills are examples of areas in which kittens learn a great deal from watching their mother. Subsequent house-soiling problems can sometimes be traced back to early learning failure due to a poor maternal example, and this is far more likely to be the case with an exclusively indoor pedigree cat than with an independent farm moggie. Farm cat offspring will also have been given an excellent example in the hunting field and lack of hunting ability is far more likely to be seen in the offspring of the cattery environment. Indeed, for people who find the murderous qualities of the cat less than desirable, this lack of hunting prowess can be a distinct advantage.

At what age was the kitten acquired?

It is not only the type of early environment that affects the adult behavioural characteristics of the cat, but also the amount of time that the kittens spend in that environment and the experiences that they are afforded. Much has been written about the importance of the socialisation period to the behavioural development of the dog, and nowadays there are puppy parties and puppy playgroups, designed to maximise the advantages of early socialisation. However, until recently far less importance has been given to the effect of early life on the adult behaviour of the domestic cat.

We know that socialisation is essential for any domesticated animal that will be expected to live in a human family and often share its life with members of other species. Kittens are known to socialise effectively with almost any other mammal with which they have close contact, provided the contact occurs when the kitten is of an appropriate age. It is only as recently as 1988 that research carried out by Karsh and Turner determined the timing of the socialisation period of the kitten and found that it occurs considerably earlier than in the puppy. Kittens were divided into several groups and then given intensive handling for periods of forty minutes each day over a four-week period. The age at which the four-week period commenced was different for each group and it was shown that those kittens handled from the end of the first to the end of the fifth week of life, and those handled from the end of the fourth to the end of the eighth week, were noticeably less sociable than those handled between two and six weeks and three and seven weeks. Combining these results with the findings that kittens which are not handled until the seventh week are no more sociable than those not handled until their fourteenth week, the researchers concluded that the optimum time for socialisation runs from the second to the seventh week of life.

Obviously nothing in life is so cut and dried, and there will be slight variations due to factors such as the individual kitten's personality and the influence of its mother. Often farm-reared cats show a tendency to be slightly nervous in human company, which results directly from a lack of very early contact with humans. Queens will often give birth in well sheltered and inaccessible places and will actively discourage contact between their kittens and the people or indeed other animals in their immediate environment.

Recently I stayed on a farm near Hereford with my family and one morning before breakfast I went for a walk round the farmyard with my son, David. As we were returning to the house, eagerly anticipating our breakfast, we saw an adult queen disappearing into an old cider press. David followed her, wanting to discover where she had gone, but found that the entrance was too small for him to see anything. All we could hear was the faint high-pitched miaowing of a litter of kittens who, the farmer later informed us, were nearly eight weeks old. They had only been seen on a handful of occasions and then only very late in the evening, when the farmyard was quiet and deserted. As they grew up these kittens would undoubtedly be extremely wary of the humans around them and any that found their way into people's homes as pets would take some considerable time to feel at ease.

One might think that the pedigree cat, reared in the cattery, would not experience the same sort of problems since it spends the early part of its life in a totally managed environment, surrounded by humans. However, cattery staff are often very busy people who have little time to spend cuddling kittens; consequently the actual level of direct human contact may be much lower than one might think. In contrast, one-off litters produced by a much loved pet and reared in the midst of a chaotic family home will usually be showered with doting human contact from the earliest days. As a result these kittens fit easily into a family of their own when the time comes to leave the nest.

What is the cat fed on and have there been any changes in its diet?
As we discussed in chapter 2, there may be direct links between diet and behaviour, and any drastic and recent changes in the cat's diet would alert my attention to this possibility. However, the evidence is not always so conspicuous, as the cases of Millie and Tabitha illustrated: it may take some careful questioning about the diet and the timing of problems in relation to feeding before a link can be established. Even then, with the absence of any detailed research in this field, it may be that a suspicion of dietary involvement can only be confirmed by trial alterations in feeding practices and the responses that they achieve.

Have there been any recent alterations in the home environment?
In the case of feline behaviour problems this is probably one of the most useful questions that can be asked. However it is also one of the most far-reaching and consequently difficult questions to answer and I often find that I have to ask it in various forms before I am satisfied that I have received a full reply.

People are only too willing to give information about any structural alterations to their home, such as extensions, or about internal alterations such as redecorating or acquiring new furniture. Both these changes may be relevant, especially if they coincide with the advent of a behaviour problem in the family cat. One client recently came to me with a cat that had started spraying in their new house. During the consultation she told me that she and her husband had bought a derelict house that they were attempting to live in whilst doing extensive restoration work. Not only had Jasper, their sensitive Burmese, had to cope with the upheaval of moving house and territory, but he now found himself living in the middle of a building site—and the owners wondered why he had started to spray!

Not all cats have such drastic changes to contend with, but alterations that seem minor to us can pose real security problems for some individuals. My own moggie found Christmas a terrible trial every year, not because of the pressures that all of us recognise at the festive season, but simply because her bean bag had to be moved to accommodate the Christmas tree!

Many people are aware of the sensitive nature of the cat and expect to see links between household changes like these and their pet's behaviour, but there is much more to the home environment than just the four walls, and personal changes can be very relevant to the cat and its problem behaviour. Cats are aware of even the most subtle changes in their core territory. Simple family tensions which are part of life for the humans in the house can have an effect on the cat that few owners recognise, and often it takes a third party such as a behaviourist to make the connection.

More major alterations to family life may also affect our feline friends. Family separations are recognised as being very traumatic for everyone concerned, but happy changes in the environment can be equally traumatic from a feline viewpoint. For an animal that lives in a scent-orientated world the arrival of a new baby, with its alien human scent and accompanying unfamiliar smell of baby equipment, can pose an enormous challenge. Similarly, visitors to the house may alter the territory scents sufficiently to affect the cat's feelings of security, and long-term guests or a new lodger may not be as irrelevant as the owner would imagine.

Does the cat have access to outdoors?

We have already established that environmental influences can affect behaviour and that cats that are kept indoors all the time may have certain resulting behaviour problems. However, there may also be external factors affecting the behaviour of those cats that are allowed to walk on the wild side. Once we have established whether or not the cat ventures outside of its core territory—the home—we have to make enquiries about the broader territory which may include the local housing estate or nearby railway embankment as well as the owner's garden. How does the cat gain access to this wider territory and what percentage of her time is spent outdoors? Is there a cat flap available, and if so do any other local cats use it to gain access to the house? Have there been any recent alterations in the local cat population? Does the cat actively defend her territory against other cats? All these questions help to contribute to an overall picture of how the individual fits into her local environment, and point to possible sources of tension that may be affecting her security.

What litter tray facilities are available to the cat?

It is undoubtedly true that the majority of behaviour problems in the feline world are related to the messy things in life and that indoor toileting and marking problems make up the lion's share of my cases. For these cats the inappropriate provision of toilet facilities, be it in terms of the litter type used or the position of the tray, is a critical factor and one that can often be easily rectified.

Trinder was a very amiable and endearing little domestic shorthair who seemed on the surface to be an ideal family pet. She was well cared for and showed absolutely no signs of compromised security. On questioning the owners I quickly identified her problem as one of house-soiling rather than marking, but the question was why? When I asked about the provision of litter trays all seemed to be in order. The type of litter was suitable and the cleaning regime more than adequate, so what was causing Trinder to look elsewhere for a latrine? The answer lay in a couple of photographs that the owners very kindly brought to show me. In the first shot Trinder was posing somewhat reluctantly with the family's Bull Mastiff dog. Being quite a fan of that particular breed, I commented on what a lovely animal the dog was and they promptly handed me a second photo, showing him lying sprawled out in his rather extensive plastic dog bed. Quite unintentionally, the owner thereby gave me the vital clue to Trinder's problem: in the corner of the photo, almost obscured from view, was a small litter tray. It is perfectly acceptable to expect our two favourite species to happily cohabit and share in the provisions of a mutual den, but I feel that we are expecting just a little too much if we ask our cats to adopt that most vulnerable of positions on their litter tray under the watchful glare of a nine stone canine!

All these questions are designed to obtain the maximum amount of information about the cat, its owners and the environment in which they live. Getting satisfactory answers will often take some time and it is the provision of adequate time that is the key to dealing with behaviour problems. It is important not only to establish the root cause of the problem, but also to help the owners understand their cat's behaviour before trying to institute treatment programmes.

Part Two

WHAT IS A CAT?

5

WILD CAT—DOMESTIC CAT

The majority of cat problems referred to pet behaviourists involve what is fundamentally normal feline behaviour in an inappropriate context. The behaviour we observe in our cats is part of their natural instinctive repertoire, and by taking a look at how cats behave in the wild we can get a deeper insight into the tiger behind our fire-loving pet. Most cat owners recognise and relish that irresistible independent streak which the majority of our domesticated cats retain. They maintain a certain aura of mystery, and when they are in a position to wander at will in the outside world we are left wondering just what they get up to when they are out and about.

As I mentioned in chapter 2, in contrast to the dog which we have genetically manipulated over the years to produce some 450 breeds, ranging from the Chihuahua to the Great Dane, the cat has largely withstood man's attempts to alter its genetic makeup. Despite the fact that the Cat Fancy recognises a wide variety of breeds, their physical appearance, from the Devon Rex to the Maine Coon, is fundamentally very similar. Behavioural studies of feral cats have shown that the domesticated cat does not differ very significantly from those living wild, and we can learn a great deal about how and why our cats react in given circumstances by paying attention to the wild cat within.

A TO Z OF FELINE BEHAVIOUR

As you read the following pages, think about your own cat and I guarantee that by the end of the chapter there will be a knowing smile on your face!

A. Agility

One striking feature of all members of the Felidae family, from the largest panther to the smallest Scottish wild cat, is their incredible grace and agility when they move. The feline body is built to climb, pounce and jump and its short bursts of speed are ideal for catching prey which is in most instances smaller than the cat itself. This is in sharp contrast to the dog which is built to cover long distances at high speed and to catch prey which is usually bigger than itself.

B. Breeding

Cats have several oestrus cycles in a year and are therefore termed polyoestrus. This is true of all members of the Felidae family, but there do seem to be differences in the reproductive cycle of cats depending on the part of the world in which they live. Cats living in tropical regions are able to have kittens or cubs at any time of the year, although there may be one season in which the majority of births take place. Those cats which live in the temperate regions, on the other hand, tend to be seasonally polyoestrus, which means that they are only in oestrus at specific times of the year. In our domestic cats the length of the day controls the season for breeding, with increasing day length triggering oestrus.

C. Chase

Chasing is a fundamental component of the feline behavioural repertoire and one that we admire in our wild cats and encourage in our pets. The ability to maintain the chase when hunting small mammals is essential if the cat is to be successful in capturing its prey. This behaviour is learnt by the kitten at a very early age. There is a great trend in our education system these days towards the concept of learning through

Chasing is a fundamental component of the feline behavioural repertoire.

play, and this is exactly the system utilised in nature. As we observe litters of kittens or tiger cubs playing and chasing each other's tails, we are actually witnessing a very detailed lesson which teaches them skills that will equip them for hunting and ultimately for survival. Rapid movement invokes an instinctive desire to chase and kill and it does not seem to matter whether the object is a mouse or a biro—if it moves then kittens chase it!

D. Digging

This is a behaviour more often associated with dogs, but it is also a feature of the feline repertoire. It is well recognised that cats are very clean creatures and digging is an important part of their eliminative behaviour. Digging a hole is the prelude to elimination and cats will then urinate or defaecate into their carefully prepared latrine before covering it over. Unlike other animals, cats do not use digging as part of their hunting behaviour, for instance to dig into rodent burrows. The main reason for this is that the cat must use its toes to loosen and move dirt because the claws are retractable and therefore are not available to do the work. In the context of the cat, digging is restricted to the raking of relatively loose dirt and this probably has some bearing on the fact that cats prefer fine material as a substrate for litter trays.

E. Eating

It is blindingly obvious that eating is a major part of the behavioural repertoire of any species, for without eating none of us would survive! What is not always so obvious is that the way in which an animal eats varies enormously from species to species, and feeding behaviour is an area of great interest to animal behaviourists. Unlike dogs, whose instinct is to polish off their food in the shortest possible time, cats have far more ability to self-regulate their food intake. No one would put down the daily food ration for their dog in the morning and expect it to eat sensibly throughout the day, but that is precisely what many people do with their cats. Obesity is by no means unheard of in the feline world, and certainly there are cats who demand food from their owners in a similar way to dogs, but in general cats view eating as a way of fulfilling their nutritional needs rather than just a very desirable way to pass the day.

F. Fouling

We have already seen that the popularity of the cat is steadily increasing in this country, and if we ask people to give a reason why they believe that cats are more suitable companions in the twentieth century, by far the most common reply will be 'because they are clean'. The problem of dog dirt on our pavements is one of the major arguments used by the anti-dog lobby, and certainly dog mess is considered by many to be the most negative aspect of dog ownership. Cats, on the other hand, are renowned for their fastidious attention to cleanliness, and kittens win a convincing victory over puppies in the house-training stakes! In the majority of households today both partners go out to work, and with the trend to delay families until later in life, many people are turning to pets to fill their emotional needs. In these situations, taking on a puppy with the house-training period ahead is simply not practical, whereas a kitten that has already learnt to use a litter tray will fit far more readily into the modern work-orientated home. Likewise adult cats, who can cater for themselves during the day and let themselves out through the cat-flap to relieve themselves, have an obvious advantage over the dog when people are increasingly staying late at the office and keeping very irregular hours.

G. Grooming

In addition to their reputation for cleanliness in their toileting habits, cats are also valued for their fanatical attention to personal hygiene. Indeed, when a cat is not bothering to wash itself we at once realise that all is not well—a reluctance to groom is a very common reason for taking the cat to the veterinary surgeon. Grooming is of vital importance to cats and they are estimated to spend some 30–50 per cent of their waking time attending to their personal appearance. This is often cited as an advantage over their canine rivals who rely so heavily on their owners to do the grooming for them.

In the feline world grooming has several purposes, the most important of which is to maintain healthy skin. Loose hair is removed, which prevents the coat from becoming matted and also reduces the incidence of skin parasite infestations. Dander or scurf is also removed. During the summer months the cat uses grooming to help in its temperature regulation—up to a third of the heat loss by evaporation can be achieved by licking

Cats spend some 30–50 per cent of their waking time attending to their personal appearance.

the coat and skin. Grooming also plays a part in the relieving of tension in cats and it is common to see them grooming furiously after an encounter with an aggressive individual or after being suddenly frightened, for example by a thunderstorm.

H. Hunting

This aspect of cat behaviour is probably the most emotive of them all. The hunting expertise of the cat is the primary reason for its domestication and yet many owners find it difficult to accept that they are housing a potential mass murderer! All members of the Felidae family hunt, but the techniques used to catch and kill their prey do differ between the species. The cat has been described as the perfect carnivore not solely because its nutritional requirements dictate that it is, but also because its entire body, from its claws and teeth to its locomotor and digestive systems, is tuned to a predatory lifestyle.

The process of learning to hunt begins in the nest at a very early age, when the queen brings back dead prey which the offspring then learn to identify using their senses of smell and taste. As time goes on the queen will start to bring back injured but live prey and the kittens will be taught to develop their own hunting skills. Although the cat's attention may be drawn to a potential victim by the noises that it makes, the primary factor in initiating the hunt seems to be the sight of moving prey. The predator will then stalk its victim until it reaches a striking range, whereupon it uses a pouncing attack to capture it. The forepaws are used to restrain the prey and position it for the kill. The behaviour patterns used in the chase are believed to be innate in all felines but the finer points of successful hunting, such as seizing and killing, need to be learned and are classed as acquired behaviour. Certainly inherited predatory tendencies are shown to be influenced by early experiences.

I. Infanticide

There are certain sectors within the cat world that believe male cats should always be kept segregated from nursing mothers and their young kittens. The basis for this opinion is the fear that tom cats might attack and indeed kill the offspring. The belief has been held for centuries and there are various references to it in literature. Over two and a half thousand years ago the historian Herodotus visited ancient Egypt where the cat was still revered as a deity. He made several observations of feline behaviour, including the infanticidal behaviour of the male, and concluded that it was the result of the tom cat's sexual obsession—that by murdering the offspring the cat was attempting to make the females return to oestrus more rapidly. This viewpoint was reiterated by Edward Topsell in his *Historie of Foure-footed Beasts*.

It is true that infanticidal behaviour has been reported in males of some of the wild felids, including lions, tigers, pumas and ocelots. Indeed, there are also cases reported involving the domestic cat which, although rare, serve to perpetuate the murderer image. In the case of lions, males which have taken over a territory that houses a fertile female and her offspring may indeed indiscriminately kill her cubs, and it is thought that such behaviour is used both to reduce the reproductive success of their rivals and to increase their own genetic survival by causing the female to return to oestrus so that they can mate with her. This shows a potential biological advantage but what, if any, can be the advantages in the case of male domestic cats who are reported to kill their own kittens and therefore eliminate their genetic progeny?

The answer seems to be none, and according to German ethologist Paul Leyhausen the infanticide of the domestic tom is not a deliberate act of murder but rather the result of a misunderstanding. A few weeks after kittening the female domestic cat sometimes exhibits a false heat (pseudo-oestrus) and the male reads such behaviour as an invitation to mate. However, the female is not receptive and she fights off the tom's advances. The tom, now in a state of sexual arousal, mistakes the low, crouched position of the kitten for the sexually receptive stance of an in-oestrus female and this misconception is supported by the fact that the kitten is unable to move away and therefore does not put up any opposition. When the tom mounts the kitten he holds it by using the neck bite which is a normal part of the mating behaviour and the kitten responds by staying very still, perhaps confusing the male's neck hold for its mother's. However, when the tom cat is unable to mate with the kitten because it is too small, his reaction is the same as to an unco-operative female and the grip on the neck becomes tighter and tighter until eventually the kitten dies.

Reports suggest that the kitten will then be eaten, but such behaviour may be triggered by a different mechanism. Dead offspring are often eaten by their parents in order to keep the nest clean and protect the remaining kittens, and thus this final act may not be related to what has gone before. Despite the fact that infanticidal behaviour does occur in the feline world for whatever reason, it cannot realistically be described as common and there are numerous reported cases of tom cats behaving paternally towards their offspring. As with so many things, it is the rare cases that have been seized on for their sensationalism and used to perpetuate a cannibal reputation for the domestic tom.

J. Jumping

The cat is recognised for its grace of movement and many a fascinating hour can be spent watching it going about its daily routine. It would be inaccurate to describe the cat as an athlete, but we all acknowledge its tremendous ability to jump either vertically onto walls and worktops, or horizontally from building to building as it patrols its territory. Whether the jump is used to reach forbidden food in the kitchen or to catch prey in the great outdoors, the starting position is basically the same. A jump onto a well-known surface may well be accomplished directly from walking, but ascent onto an unknown surface or up a great height will be carefully measured and assessed by eye. The cat then crouches slightly as the bodyweight is positioned over the hindlegs, and these are then rapidly extended to propel the cat forward. The propulsion during the jump involves tremendous levels of energy, far in excess of those required for other locomotor activities. A good jump will propel the cat slightly above the objective, so that there is room for it to land on its back legs while the front legs are used to correct the balance.

When it comes to a descending jump, however, the cat is far less elegant. By stretching the body down as far as possible before take-off, it shortens the jump and also lessens the impact on the front legs as it lands. It will land with its forelimbs, placing them on the ground one at a time, the second in front of the first. The back legs are then rapidly brought down to spread the load. When the descent is from an exceptional height the cat will often make an intermediate jump which enables it to use its back legs to push away from the wall and convert the downward jump into a horizontal leap.

K. Killing

As we have already discussed, hunting is a major component of the feline behavioural repertoire, and it is indeed a hunt to kill. In general small prey is killed by using the nape bite. This is designed to dislocate the cervical vertebrae and it is believed that nerve receptors at the base of the canine teeth are used to direct them to the correct location for a fast and effective kill. In the case of larger prey the throat bite is usually used. This works by blocking the trachea or windpipe, so leading to the victim suffocating. This method of killing is usually only associated with the larger cats, since these tend to catch the larger varieties of prey. To a great extent, the method used to kill the prey is dictated by the species of cat involved and by the sort of prey, but individual preference is also a factor and it is possible that the killing technique may be altered if the prey proves difficult to kill.

For the domestic cat the nape bite is the culmination of the hunting process and the one aspect that the kitten is unable to perfect during play. Practice on prey is essential, and the kittens will accompany their mothers on hunting trips in order to improve their skills. Inexperienced kittens often fail to use enough strength to kill their victim, but competition between littermates increases the excitement level and with repeated practice they master the technique. Occasionally the bite is not directed accurately: the kitten takes hold of its prey by another part of its body and the prey may turn round and bite the kitten. The first kill that a kitten makes is usually accidental, and not until it has made several successful kills will it learn to associate the neck bite with its success.

L. Language

When we talk about language we usually mean the verbal communication that is at the centre of our human interactions. Language in the feline world, however, is far more complex and consists of a combination of vocal and non-vocal components. Cats certainly use sounds to convey information, and research has identified six basic calls that are characteristic of small wild cats—the hiss, spit, purr, growl, miaow and scream. The usage of these six sounds varies in any given circumstance, and one sound may be changed to another via a combined call during the course of a particular encounter.

In the case of the domestic cat, scientists have distinguished sixteen different sounds and it is likely that the cat could identify numerous others besides. Some calls can easily be linked to actual meanings—the snarls and shrieks heard during the course of a cat fight are fairly unambiguous, and the deep throated growl that my moggie frequently utters when confronting my rather infuriating colourpoint needs little translation! Other readily identifiable sounds include the food- or attention-demanding miaow, the distress howl, the welcome home trill and the oestrus call.

All these sounds can be categorised into one of three groups depending on the way they are produced, which Michael Fox labels as murmurs, vowels and high-intensity sounds. The murmurs are made with the mouth shut and include the soft sounds such as the greeting call and the purr (I shall deal with the purr later under P). The second category, the vowels, includes those sounds used by the cat when communicating with its owner, and they are made by gradually closing the mouth. These are perhaps the most individual of the vocal communications made by the cat, and those of us who live in multi-cat households have no problem distinguishing between the vowel sounds of the different cats. High-intensity sounds are generated by keeping the mouth open and altering its shape to produce individual sounds. The growl, hiss, spit and shriek are all produced in this way and in general high-intensity sounds are those used in communication with other cats. So as we can see, there is an extensive vocabulary in the feline world and vocal language has an important part to play in communication.

M. Mating

The mating behaviour of all Felidae is essentially very similar. All the species engage in prolonged courtship rituals which seem to be designed to reassure both the male and the female that their intentions are mutual in order to prevent a misunderstanding which could result in physical violence. Both visual and vocal communication is used in combination with the important scent messages. In the case of wild cats the male and the female may spend considerable periods of time together before mating and the female often acts provocatively towards the male while at the same time resisting his attempts to mate. It is thought that she uses this time to make certain that the male is the owner of the territory and not some intruder who will eventually be driven away from the area by the resident male.

With domestic cats the courtship period is somewhat shorter but is just as complex. The main factors affecting the length of the courtship are the breeding experience of the male and the familiarity of the place where mating is to take place. Increased experience and increased familiarity will both result in less time being spent in the courtship routines.

Once the male has approached the female he will sniff at her genital region and exhibit what is called a *flehmen* reaction (see under Odour, p. 57) in which he extends his head, neck and upper lip and then grimaces. This behaviour is associated with the use of a small scent organ called the vomeronasal or Jacobson's organ which enables cats, and other species that possess them, such as horses, in effect to taste the smells that they encounter. By using this organ the tom cat is able to detect the sex pheromones produced by the female. These are chemical substances that communicate information between individuals and affect their behaviour.

As the male approaches the female she assumes a receptive position with the pelvic region elevated and the tail held to one side while she treads with her back legs. The tom will then proceed to mount the queen and take a neck grip on her with his teeth. This grip is not a form of aggression since the male cat is actually inhibited from showing aggression to an in-oestrus female. Its purpose is to immobilise the queen and so protect the male from any attack she may launch upon him. It is not a case of brute force used to squash the female into submission but rather a freeze reaction in response to being held by the scruff of the neck, in the same way that queens immobilise their kittens by grabbing them firmly by the scruff.

Initially the tom cat positions himself fairly high up on the female's back, but then he starts to tread with his back legs and gradually moves himself backwards until he is in the right position for copulation. The treading by the female also helps to achieve the correct positioning. Intromission is accompanied by pelvic thrusting and the tom remains motionless for a few seconds before releasing his neck grip and rapidly dismounting. The queen lets out a loud cry and swipes out at the tom before licking her genital region. She then goes into the 'after-reaction', rolling and rubbing on the floor. This behaviour is thought to be due to the short, sharp spines that cover the surface of the tom cat's penis and are used to trigger ovulation in the queen who is what is known as an induced ovulator. Insertion of the penis is not painful, but when the tom withdraws following copulation the queen reacts with an ear-piercing cry. Some people disagree with this explanation and say that the fact that the queen is receptive to mating again so quickly after the tom cat's withdrawal refutes the idea that there is any pain involved. It has been suggested that her violent reaction is a defence mechanism in response to the fear that the male might become vicious towards her once the mating process is complete. The German behaviourist Paul Leyhausen found that a wild female cat kept in captivity would not scream or attack the male following mating, provided that she knew him well, and this seems to support the theory of a defence reaction.

N. Nursing

Once the kittening process is complete the queen will stay with her kittens almost continuously, rarely leaving them for more than two hours at a time. When she does leave the nest she does so only to feed, relieve herself or exercise. Nursing may not start for up to two hours after kittening but once it does the queen will spend some 70 per cent of her time during the first week nursing her young. For the first three weeks she initiates all the nursing sessions, licking the kittens to arouse them and directing the licking to help orientate the blind youngsters towards her teats. She lies on her side with her body arched around the litter and her teats exposed. With time the kittens are able to orientate themselves to find the teats, and by the second or third day many kittens are able regularly and repeatedly to take up specific nipple positions. When the queen lies down and presents herself all the litter will feed although not all kittens will nurse at once. The queen needs to lick the anogenital region to stimulate the kittens to urinate and defaecate, and she is therefore always able to clean up all the waste as soon as it is produced. This prevents the nest area from becoming soiled.

After three weeks the kittens are able to see and hear and can leave the nest to explore their surroundings. Nursing sessions are now initiated by the kittens and can take place either inside or outside the nest. The queen will usually respond by immediately lying down and exposing her nipples for them. As time progresses she begins to avoid the nursing sessions and will lie so that her nipples are hidden or will climb up to places where the kittens cannot reach her. As she becomes less and less available to the young and they become capable of taking adult food, the weaning process becomes complete. Towards the end of this period the queen will often bring back rodents to the nest, and so the kittens' hunting education begins.

O. Odour

For the feline members of our society it is an odour-centred world. The sense of smell is a vital one for the cat and it is utilised extensively in many areas of behaviour, from feeding to territory marking and from aggressive displays to sexual interactions. It is difficult for us humans, who rely so heavily on visual senses, to appreciate the subtleties of the odours around us and to identify with a creature that approaches every novel stimulus with its nose!

The sense of smell, or olfaction as it is scientifically described, is a chemical one and is closely related to the sense of taste. These two senses are used to monitor the chemical signals which are presented both within and outside the body. In order for the chemicals to be identified special receptor cells or chemoreceptors convey vital information to the brain. There are two very distinct sets of chemoreceptors responsible for the senses of smell and taste, but they are closely related since the nasal passages that house the smell receptors open directly into the mouth where the taste receptors are situated.

For the cat the story does not stop there because they possess a third chemical sense which is often described as a combination of taste and smell. This sense has its own special receptor organ which we have already seen the tom cat using in mating—the vomeronasal or Jacobson's organ. Situated in the hard palate, this long cigar-shaped sac connects to the mouth via a narrow duct that opens just behind the upper incisor teeth. It is lined with olfactory cells and is stimulated by chemical substances which have first been captured by the tongue and then transferred to the opening of the vomeronasal organ by pressing the tongue against the roof of the mouth. Stimulation of this special receptor organ is accompanied by the very distinctive behaviour known as the *flehmen* reaction, from a German word for which there is no English translation. It describes a very distinctive facial gesture: the cat stretches its neck, opens its mouth, wrinkles its nose and then curls back its lips in what almost resembles a snarl. The *flehmen* reaction is seen in kittens as young as six weeks and is a behaviour primarily, although not exclusively, associated with males. It is most frequently provoked by the urine of an in-oestrus female and is also seen when the tom cat investigates the genital area of his potential mate.

P. Purring

The purr is probably the most commonly recognised aspect of all cat language and is certainly the one most appreciated by the cat lover. It is thought to signify a deep level of contentment and people find it comforting and relaxing. Of course contented cats do purr, but this is by no means the sole trigger for this distinctively feline behaviour. Cats that are injured and in great pain will often purr, as will those that are seriously debilitated or even dying. When cats are brought to me in the surgery, I frequently have difficulty in hearing the heart and lung sounds via the stethoscope because they are drowned out by the incessant purring of what is a far from contented individual.

One theory offered to explain these varied uses of the purr in feline communication is that it is a friendly social signal which can be used either to indicate a need for help and a willingness to accept it or to portray a sense of contented oneness with the world. Purring first occurs when the kittens are only a matter of a few days old and is used to signal to the queen that all is well and that the kittens are suckling successfully. In turn the queen purrs to the kittens to increase their feeling of security and to communicate her relaxed and co-operative disposition. Later on the purr is used by the queen as she approaches the nest and the kittens will purr as they approach humans or adult cats with whom they wish to play.

The purr is remarkable not only in the way it is used to convey a variety of messages but also in the way it can be produced with such continuity. A cat can purr for considerable periods of time without a change in either the intensity or the rhythm, and it can be continued through both inhalation and exhalation, even with the mouth closed. This two-way purr is exclusive to the domestic feline species: the big cats such as the tiger are only able to purr with each outward breath. In this respect the domestic cat may be superior to the big cats, but they compensate for that by the fact that they are able to roar, which small cats cannot do.

Purring is certainly a fascinating method of communication, but despite the fact that we recognise it so readily we are still unable to explain exactly how it occurs. The most frequently offered explanation is that it is produced by the vibration of the false vocal cords, two folds of membrane situated behind the true vocal cords within the larynx. However, this theory is not accepted by everyone and there are those who suggest that it is out-of-phase contractions of the larynx and the diaphragm that result in the purr. A third theory involves turbulence in the bloodstream of the main vessel that returns blood to the heart, that is the vena cava. It is thought that when the cat arches its back, the blood within the vena cava forms eddies at the bottleneck which has been created where the blood vessel constricts to pass the liver and the diaphragm. This then sets up vibrations within the thorax or chest, which are passed up the windpipe and resonate in the sinus cavities of the skull. Whatever the true explanation, one thing is certain: the purr that we all find so comforting is one area of feline behaviour that warrants a great deal more investigation.

Q. Quarrelling

When we talk of quarrelling in the context of the animal kingdom we all too readily think of physical conflict and out and out aggression. The truth is that by far the majority of quarrels within the feline world are settled without resorting to violence, and when you look at the efficient weapons that every cat posseses in the shape of its teeth and claws, it is easy to see why this is an advantageous situation. Cats do not live in mutually dependent packs in the same way as dogs and they do not depend on a stable hierarchical regime. Hunting is a solitary occupation in most contexts and cats are therefore keen to avoid conflict in order to minimise the risks of injuries that might prevent them from hunting and ultimately surviving. Even a very confident and 'dominant' individual will put himself at risk from the claws of his opponent if he attacks, and so he prefers to settle the disagreement via complex displays of body movements, facial expressions and vocalisation. Such displays are regularly encountered in the multi-cat household where cats that live together develop what owners describe as a 'healthy respect' for one another. In these circumstances the sight of the cats staring each other out across the kitchen or the sound of the deep warning growl may be commonplace, but physical violence and injury are exceedingly rare.

R. Raking

The cat's cleanliness with regard to toileting is the quality that scores it most points over its rival the dog. The hygiene

The majority of quarrels are settled without resorting to violence.

and health implications of dog faeces and indeed urine are a constant source of debate and disagreement. The cat, on the other hand, is renowned for its fastidious cleanliness and the subject of cat faeces is far less emotive for most people—apart, perhaps, from keen gardeners who have the misfortune to live next door to a cat.

Although it may be nice to believe that cats cover over their excreta in order to please us and make our lives more pleasant, this is obviously not the case. Raking over to cover up both faeces and urine is an important part of normal feline behaviour and part of the subtle use of odours in feline communication. It would appear that the behaviour is triggered simply by the presence of excreta, but there are situations when the cat will deliberately fail to cover his deposits and rather leave them prominently displayed. Such a use of faeces is termed 'middening' and is most likely to occur at the boundaries of adjoining territories.

The urge to rake over can be very strong, and some cats

will continue to rake long after the faeces are obscured from view. Others rake over the area adjacent to the litter tray as well as the litter itself, and this may even be extended to raking nearby objects that are quite different in both appearance and texture from the litter substrate.

If we watch a cat while it is raking over its most recent deposit we see it stop every so often to sniff the area and presumably smell the covered faeces. This would suggest that the cat continues to rake until the odour emitted from the excreta reaches the desired level, and we can assume that the level will vary in different circumstances. If the cat does not wish to be detected it will rake over until the odour is completely disguised, but if it wishes to identify itself with the territory it will rake until sufficient odour persists through the substrate to communicate to the other cats in the area. Unfortunately we cannot appreciate the subtleties of the smell of cat faeces and so we shall probably never be able fully to understand the importance of raking. However, we can continue to appreciate the cleanliness that results from it!

S. Scratching

Scratching is a complex behaviour that has a multitude of uses. It is one aspect of feline behaviour that does not endear cats to their human owners and is a frequent source of contention. It is also a much misinterpreted behaviour, many owners believing that their cat is merely sharpening its claws. The cat does indeed scratch to 'sharpen' its claws but not in the sense that most people think. It would probably be more accurate to describe the process as 'conditioning' the claws, for what actually happens is that the frayed and worn outer claws of the front feet are removed by scratching, to expose the new, very sharp claws which are already growing beneath. The worn-out claws are often found at the base of the scratching post. In the case of the hind feet the worn claws are removed by using the teeth to chew them off.

The second function of scratching is to exercise and strengthen the apparatus used to protract and retract the claws, which is vital for cats in fighting, climbing and catching prey.

The third function, and the one most often overlooked, is that of marking. Scratch posts serve as a visual marker in a cat's territory and at the same time they act as scent marks.

On the underside of the cat's paws there are small interdigital scent glands, and the rhythmic stropping of the front feet along the scratching post activates these glands to release their scent. At the same time the sweat glands on the pads are activated and the cocktail of secretions from these two types of glands result in a scent mark which is unique to the individual.

T. Territorial

For many cat owners the frequent visits to the vet for treatment of cat fight abscesses are adequate proof that the cat is a territorial creature. Certainly duels between cats in adjoining or overlapping territories are common but we tend to be unaware of the countless other cases where cats have come to share their patches peacefully or even amalgamate them. Every cat has its own home base which is surrounded by the home range and beyond that the hunting territory. In the case of the wild or feral cats the territory boundaries are set up according to the state of the local cat population and the availability of food. For the domestic cat, however, the territory is decided for them to some extent by their owners who are usually completely unaware of the existing position as regards feline territories in the area.

The home base is usually considered to be the actual house in which the cat lives, and for some cats in multi-cat households it is reduced even further to merely a particular chair or resting place. The home range usually includes favourite places for playing, sleeping, dozing and sunbathing and the extent of it is determined by various factors, including the number of cats in the area, the food supply and the sex, age and personality of the cat and its surrounding neighbours. In multi-cat households the home range is a shared responsibility and will tend to be larger than that of single cats, but in either case it is unlikely to be any larger than the garden that surrounds the house.

Females and neutered cats tend to occupy small but well-defined home ranges and they will usually defend these with vigour. Entire tom cats have larger, more diffuse territories which in some cases may be as much as ten times the size of those controlled by the females. The borders of these territories are less well-defined and the toms tend not to defend them so single-mindedly. However, when a strange cat does enter the range, fights will occur.

Beyond the home range lies the hunting ground and this is connected to the home range by specific routes. These long, circuitous paths go round the neighbouring territories which are defended by other cats. The areas between the paths are rarely used and cats are careful to avoid meeting each other, so ensuring that it is seldom necessary for fights to occur. A cat which is about to set off on one of these shared routes will check the path ahead and wait till all is clear. Time-scheduling appears to play an important role in the feline world, and cats that share territory boundaries may establish routines whereby one cat has the right of way in the morning while the other has the right of way in the evening. (Such time-sharing has also been identified in multi-cat households within the home range.)

When meetings do occur the cat which is already on the path appears to have the right of way, regardless of its social standing, and vocal and visual signals are usually enough to resolve the situation. However, there will be occasions when neither party will give way and fighting may occur. The interesting thing is that these conflicts, when they happen, do not confer permanent status on the winner; it is just as likely that a future conflict between the same individuals will have the opposite result.

U. Urine Spraying

Scent marking is a fundamental behaviour of all Felidae and is recognised in species ranging from the serval to the bobcat and from the tiger to the domestic moggie. It is a powerful method of communication, and although it is more commonly associated with the male of the species it is in fact practised by females and neuters as well. The reason why we attribute it to the males primarily is probably due to the fact that male feline urine is so pungent that the marks created by it are easily recognised, even by the inferior human sense of smell. Female and neuter urine, on the other hand, may go undetected.

The spraying of urine is a deliberate act and must not be confused with urination in response to a full bladder. It has been shown that cats will continue to spray in a set routine regardless of the state of their bladder, and the area they spray and the number of squirts they perform will remain constant despite fluctuations in the amount of liquid they consume.

Another fundamental difference between spraying and urination is the posture the cat assumes during the act. Urination is performed in a squatting posture while spraying is performed from a standing position with the cat facing away from the object to be marked. The cat will stand a few centimetres away from a tree, shrub, fence post or other vertical object and, holding its tail high, will direct a small volume of urine backwards in short squirts that result in the urine being spread over the target area. Usually the tip of the cat's tail will quiver as it sprays, and at the same time it will often step with its hind legs, arch its back and wear a look of extreme concentration on its face.

The reasons why cats perform their urine spraying against vertical objects and from a standing position is so that the urine is deposited at nose-height for other cats. The spray acts as a personal identity card, and far from being simply an unpleasant and off-putting smell, it carries all sorts of information about the depositor. It is believed that the mark conveys information about the cat's age, sex, health status and rank and that the freshness of the urine is used to indicate how long ago the sprayer was in that particular location. This last function is thought to be important in the time-sharing system explained under Territorial, p. 62.

It has been suggested that urine spraying is the mark of an over-confident individual and that the odour is a threat signal to rival cats. Observation would appear to refute this and certainly in the tiger it is noted that urine spraying is increased in unfamiliar surroundings, supporting a theory that these animals are comforting themselves by spreading around their own distinctive odour. The reaction of cats to the spray marks of others certainly does not appear to be one of fear. In fact the opposite would seem to be the case, with cats being positively attracted to scent marks and sniffing at them with great interest. Many cats will sniff at a sprayed area and then simply walk away without any apparent effect on their behaviour, others may be induced to spray over the mark with their own urine. What is certain is that urine spraying is a very important component of normal feline behaviour.

V. Visual communication

Konrad Lorenz once stated that 'there are few animals that display their mood via facial expressions as distinctly as the cat', and we only need to observe encounters between the neighbourhood cats to realise just how important body language is in the feline world. Lions are known to use a diversity of visual signals in their communication and the German behaviourist Paul Leyhausen noted a similar use of visual signals in a wide variety of wild cats when kept in captivity.

There are various facial expressions of the domestic cat that we humans find totally unambiguous, such as the flattened ears, snarling face and lashing tail of a cat that is not best pleased with the situation! However, there are also more subtle examples of visual communication when the facial signals are combined in such a way as to be difficult to distinguish. Presumably these cats are hedging their bets and by disguising their true emotions are delaying the need to make a decision about how to react.

The ears and the tails are the most expressive parts of the cat's body and thrashing the tail is often the very first sign of anger. Tail wagging in cats is recognised as a sign of conflict, with horizontal, rhythmic movements of the tail beginning in a gentle fashion and becoming more and more pronounced as the cat is increasingly provoked. The position of the tail also conveys information to those around: a tail that is held high indicates that the cat is active and greeting, while a tail that is arched downwards indicates an aggressive state of mind. The ears are an excellent indicator of the cat's emotions, and by combining ear positions with different whisker positions and levels of pupillary constriction and dilation the cat is readily able to convey information to those around him.

It is thought that the cat uses nine clearly recognisable facial expressions and combines these with some sixteen distinct tail and body postures. In common with scent communication, used so extensively by the cat, visual communication is also used to defuse moments of conflict and avoid out-and-out displays of aggression.

W. Walking

Cats are designed as hunters and while they usually move at what can be described as a leisurely pace they are also capable of extremely slow and controlled movements which can be essential in the hunting of their prey. The walk is a very stable gait for the cat and when they need to increase their speed there is a noticeable decrease in both elegance and efficiency. Walking is described as a four-beat gait, which means that all four feet make contact with the ground at separate times. During any one phase of the walk at least two feet will be in contact with the ground, and in the slowest walk, the stalk, three or even all four.

The sequence of movement in the walking cat is usually the right hindleg and then the right foreleg followed by the left hindleg and the left foreleg. The object of this is to provide sufficient stability for the cat to stop at any moment and not fall over, and indeed it is remarkable to watch a cat which stops in mid-stalk and then proceeds to hold its position for a considerable period of time. The strong hindlegs are responsible for the propulsion while the forelegs play a more supportive role, and by swinging the front legs inward as it walks the cat is able to place its feet neatly in front of one another so that the resulting tracks make a virtually straight line. Even the hindlegs are swung inward to some extent and the result is a creature that, aided also by its narrow chest and balancing tail, can quite happily walk along fence rails and narrow ledges.

Slow movement and stalking may be important in the hunting down of prey, but the cat must also be capable of short bursts of high speed and the feline foot conformation helps in this respect. Rather than placing the sole of the foot on the ground as we do, the cat places only the ball of the foot, while the parts equivalent to the human ankle and heel are elevated. Compared with animals that are a similar size but walk on their soles, the cat has thinner and lighter bones as well as shorter and narrower feet. All these features are part of the specialisation for swift locomotion.

X. X-breeds

Unlike the situation in the dog world where the pedigree dog is very prevalent, the vast majority of cats kept as pets in this country are X-breeds—good old moggies! However, as the popularity of the cat increases, the number of pedigree cats seems to be on the increase and with that there appears to be a dramatic surge in the number of cats being kept exclusively indoors. It would seem that this is also bringing an increase in the cases of behaviour problems in the feline world, as will be seen in Part Three. Behaviour characteristics vary considerably between the pedigree cats and it is becoming possible to identify breed tendencies towards certain categories of problems in a similar way to that experienced in the dog world.

Y. Yowling

A yowl is described in the dictionary as the loud, long cry of a cat in distress, and the term is often used to describe the plaintiff cry of a cat that is trapped somewhere and unable to escape. For the domestic cat it often follows an inquisitive exploration which has resulted in the cat getting shut in a garden shed or in the next-door garage. The yowl is a distinctive cry for help which travels well and is therefore very efficient for getting the owner's attention and leading to a speedy rescue.

Z. Zoonoses

This is the term used to describe diseases that can be passed from one species to another—from cat to man and from man to cat. Probably the one that most people think of is rabies, but there are other more common examples such as toxoplasmosis (see p. 87). It is not relevant in this book to list all the examples, but suffice it to say that disease transmission between our two species is possible.

After reading this A to Z of normal feline behaviour you will be able to recognise in the letters that follow that many cases of behavioural 'problems' in fact involve quite natural feline behaviour patterns. My job as a behaviourist is first to identify whether the cat's behaviour is abnormal. If it is, then treatment needs to be started, and if it is normal then I need to establish why it is presenting a problem for the owners. What constitutes a problem is often dependent on the owner, but if he or she is concerned enough to have approached a behaviourist then the relationship between cat and owner is obviously not ideal. It is interesting to note that cat owners are generally willing to tolerate much greater disruption of their lives, for example in terms of household hygiene, than most dog owners, and whereas owners of dogs with behaviour problems tend to blame the dog, by far the majority of cat owners will blame the situation on themselves and suffer tremendous feelings of guilt and failure when their relationship with their cat is marred. It is therefore important that the behaviourist approaches the case with sensitivity and that practical advice is offered. The following pages aim to do just that.

Part Three

WHY DOES MY CAT . . . ?

Cats are arguably the most fascinating of our domestic pets, and with the swing away from dog ownership which is undoubtedly taking place in the wake of the 'Dangerous Dogs Act' and all the accompanying bad press for dogs, cats are set to become the most numerically popular pets in our land. From a behavioural therapy point of view cats present far fewer problems to their owners than dogs, and certainly at present the majority of problems presented to pet behaviourists are of the canine variety. However, I have no doubt that, with the cat's rapid rise in popularity, the number of feline behaviour problems is also set to rise. We may not see the same level of aggression or destruction problems in cats as we do in dogs, but nevertheless, the problems that cats present to their owners may be equally distressing.

In this part of the book we shall look at a variety of specific problems and how they can be dealt with. The questions are taken from the numerous owners who want to know more about why their cat behaves as it does, and all names have been changed to protect the innocent. There are some cases where the presenting problems may be very different but the treatment is the same, and this is because it is the cause and not the manifestation which determines treatment programmes. In some cases the questions are really more a query about feline behaviour rather than an actual problem, but in all cases I hope that the answers given will help you to achieve a deeper understanding of your feline friend.

WHY DOES MY CAT . . .

A

Aggression

This is a subject that is more readily associated with the dog, and recent media attention has helped to identify it as an undesirable canine quality. However, we know that dogs do not hold the copyright on such behaviour, and indeed, the human species is capable of dramatic displays of aggression in a variety of contexts. So too the cat engages in aggressive displays and at times owners who find that such behaviour is marring their relationship with their pet will come to the behaviourist for advice. Within a behavioural context aggression can be subdivided into categories according to the perceived cause—for example, learned aggression, play-elicited aggression, redirected aggression and so forth. These will all be dealt with in the appropriate alphabetical sections, but firstly we need to take a look at what we mean by aggression and how it fits into the feline behavioural repertoire.

According to a dictionary definition, aggression is an unprovoked attack or an assault, but surely this is taking a far too simplistic view of what is after all an essential facet of every animal's behaviour. There is in any species a range of normal aggressive responses which are necessary for survival, and the cat is no exception. However, when owners find themselves faced with the impressive feline weaponry of claws and teeth, few will take time to ponder over the cause of the outburst, or indeed try to put the display in context with normal behaviour. What matters is that there is a potential for injury and the effects of a confrontation with an angry feline can be dramatic and painful.

The normal human reaction is to counter aggression with aggression, and the belief that 'I am bigger therefore I will win' drives many people to try to control the situation with force. Not only is this a futile exercise when faced with an opponent of such agility and intelligence, but it is also a dangerously provocative one and often leads to unnecessary conflict and injury. It is essential never to use aggression as a 'treatment' but rather to differentiate between causes and, by taking a detailed look at when, why and how the aggression is manifested, to start treatment programmes geared to the individual.

Aggression is in most instances a defensive action, and for the cat, which is essentially a solitary predator, it is most commonly used in disputes over food sources and territory boundaries. It is not in the cat's interest to engage in unnecessary outbursts of aggression, since any conflict brings with it the risk of injury which could in turn jeopardise the animal's survival. Without a pack to protect it the independent feline relies on its own ability to hunt, and any threat to that ability is avoided if at all possible. Instead of jumping in with teeth glistening the cat engages in elaborate displays of body language and vocalisation designed to defuse the situation and remove the need for actual fighting. Such displays are fascinating to observe and most of us who live in relatively highly populated areas will have witnessed confrontations of this sort between our own cats and those of our immediate neighbours.

Recently the people who live next to us acquired a new cat, and following a period of confinement in his new home he began to venture out into the garden and put his mark on his new territory. On entering our garden he was greeted by our resident colourpoint who, despite an extremely laid-back attitude to life in general, proceeded to engage in a most impressive display which lasted for some three-quarters of an hour. From a strategic position on the patio my cat used the feline tricks of fluffing out his coat, arching his back and sitting sideways on to make the most of his stature. Adopting a crouching position, he let out a succession of low-pitched warning growls and stared his opponent directly in the eye. With a series of subtle variations in body posture and a range of facial expressions he conveyed the unmistakable message to the next-door cat that this garden was already occupied.

The intruder retreated quietly and the potential conflict was resolved without the need for actual fighting. So it is in the majority of feline encounters: out-and-out aggression is very much a last resort. As with any species, a cat's defence mechanism incorporates the three Fs—Fight, Flight and Freeze. If a cat is challenged it may momentarily freeze, but given an easy escape route most will take the flight option and rapidly remove themselves from the confrontation. It is usually only when such avoiding action is blocked in some way that the claws and teeth really come into play, and if cats are resorting more readily to violence we need to ask ourselves why their tolerance level is decreased. In some cases it may be readily apparent, but in many more there will be a subtle interaction of factors such as early experience, illness, compromised competence and even pain, all contributing to the cat's behaviour.

Agoraphobia

One of the hallmarks of the cat as an independent creature is its ability to combine a home life with an existence in the great outdoors. We have seen a marked increase in the number of cats confined totally to barracks and never allowed the opportunity to explore their wider territory, but most people assume that, given the chance, all cats would like to walk on the wild side. It is therefore difficult to imagine a cat that is reluctant to venture out, let alone one that is actually frightened to do so. Agoraphobia is defined as the fear of open spaces and is widely taken to mean a fear of the outdoors, but it can also involve fear of open spaces within a room. According to some leading behaviourists this is the only true phobia seen in the cat, and whilst it is undoubtedly a behaviour problem, it must be said that it is relatively rare. The cause of the problem may be traced back to kittenhood, resulting from either a total lack of exposure to outdoors or else delayed exposure such that the kitten had already passed the exploratory phases of development.

The fact that cat breeders do not sell kittens until after they have had all necessary vaccinations means that the majority of pedigree cats do not venture into the big wide world until they are at least twelve weeks old. Obviously it is important to protect kittens from the major feline diseases, but by exposing them to early handling and a wide range of challenges within the breeding environment they can develop the exploratory behaviour patterns that are so necessary when facing the outdoors for the first time. If delayed exposure is compounded by a lack of novel stimuli in the kittens' early environment they are likely to develop into individuals with severely compromised exploratory behaviour patterns, and therefore face the prospect of being incompetent or even agoraphobic as a result.

Not all cases of agoraphobia are rooted in kittenhood, and in fact it is more likely that the fear stems from a traumatic incident which has resulted in a total loss of self-confidence. Such incidents may involve disruption of the boundaries of home due to the building of an extension, or a frightening encounter with a neighbour's dog. Near misses with cars are often cited as triggers for a fear of outdoors, but in the feline context the most common form of single traumatic incident is an encounter with another cat. Fights with a particularly territorially aware individual may often result in the loss of more than the odd ear, and less competent individuals may emerge from such conflicts in emotional tatters.

Since venturing outside brings with it the constant risk of encountering the perpetrator of the crime, the defeated cat will become increasingly reluctant to do so. As time goes by the cat's phobia becomes more generalised, so that he is unable to cope with any external environmental stimuli. Eventually the sound of the wind in the trees is enough to send him into sheer panic, and even if the other cat is nowhere to be seen he will refuse to leave the safety of the house. In some cases, where the original attack occurred actually inside the cat's own home, the emotional effects can be devastating: such cats may develop not only a fear of the outdoors but also a reluctance to move out into the centre of a room. In most cases the phobia becomes progressively worse and these cats become visibly distressed when forced outdoors.

Question

I have recently taken in a stray cat whom I have christened Boomer. He was found abandoned under a hedge at the side of a busy road and taken to the local vet where he was treated for minor cuts and bruises before being rehomed. He is settling very well into our household, and although I may be biased, he really is a lovely cat and is the most affectionate individual you could wish to meet. However, if I open the back door he runs and hides under the table, and on occasions when I have tried to push him out he has gone into a complete panic and badly scratched me in his frantic attempts to get back inside the house. I have no real objections to him being a house cat, but I feel that he is missing out on all the fun of being independent.

Answer

In Boomer's case it is impossible to know what went on in his early kittenhood and we can only speculate as to the reasons for his apparent abandonment. However, the fact that he was found close to a road and that he needed treatment for cuts and bruises does make me wonder about the possibility of a road traffic accident. If Boomer did have a narrow escape with a car, it would explain why he was hiding under the hedge and it would certainly explain his reluctance now to venture outdoors.

Treatment for Boomer, and indeed all cats suffering from agoraphobia, involves decreasing their fear of outdoors by a process called systematic desensitisation. This involves using a large secure pen, similar to those used by breeders for kittening, which can become a safe haven for the cat, offering it much-needed protection during the early stages of exposure. The pen is placed in the garden and the cat is made to spend a gradually increasing amount of time in it each day. It may help to divide the cat's food into a number of small meals and feed these inside the pen, thereby creating pleasant associations with outdoors and also distracting the cat when it is feeling especially vulnerable.

In cases where the agoraphobia is traced to a conflict with a particular cat, it is obviously essential to ensure that the aggressor is not likely to enter your garden during the treatment programme! As with any phobia, exposure to the full-blown fear stimulus during treatment will severely compromise the possibility of success, and although some believe in throwing people in at the deep end to overcome a fear of water, you have to remember that there is always the risk that they might just drown. In Boomer's case the only proviso to the use of a pen in this way would be that it should not be put in the front garden next to the road, but I hope that that goes without saying!

Once Boomer has accepted that the outdoors is not intrinsically threatening, he should be taken out into the garden without the protection of the pen. This is best achieved by putting him on a lead so that the owner can act as a security bridge

Even if the other cat is nowhere to be seen he will refuse to leave the safety of the house.

and help to bolster his confidence. Eventually Boomer can be encouraged to go out alone and it may help to continue with the small frequent meals and put his feed bowl just outside the back door. The prognosis for treatment of agoraphobic cats is usually good, so long as we can achieve reasonable control over the original cause. Where the cat was once happy to wander outside we are merely re-establishing a behaviour pattern and these cases are likely to be easier than those where you are teaching an adult cat to accept the great outdoors for the first time.

It is unlikely that Boomer will ever be happy to venture far beyond the garden and the sound of traffic may continue to instil fear for some time to come, but many owners see this as a distinct advantage if it keeps their precious pet off the roads.

Anorexia

Anorexia is a medical term that quite simply means the decrease or cessation of normal feeding behaviour, and often it is a change in feeding behaviour that first alerts an owner to the possibility of a problem. Any cat that loses its appetite needs to be investigated since there are numerous medical reasons for a lack of interest in food. These range from serious illnesses to sore gums and bad teeth and a visit to the vet will identify the problem.

If it is the teeth that are the cause of the problem, the cat will usually appear to be keen to eat but back away from the bowl after only a couple of mouthfuls because of the often quite severe pain that results. Veterinary surgeons nowadays offer a comprehensive dental service to their patients and it is always worth asking your vet to take a look at your cat's teeth whenever you visit the surgery. As with so many things, prevention is better than cure, and with a little attention to diet and dental hygiene most dental problems can be avoided.

Occasionally owners report that their cat will starve itself rather than eat any food other than its favourite variety, and certainly there are some particularly finicky cats about. These cats are not usually suffering from any medical complaint but rather have learnt that by refusing to eat they can force their owners to give them a more tasty option for their tea! I am always amazed at the vast numbers of cat owners who daily cook a meal of fresh meat or fish for their beloved feline and at those who actually hand-feed their moggies because 'he doesn't like to eat from the bowl'!

The situation is much the same as the three-year-old child who refuses to eat anything other than fish fingers and baked beans, and the treatment is much the same too. The owner must stand her ground and remain firm, but for owners of some cats that have become addicted to one particular food, making the change can be far from easy.

Feeding a healthy and balanced diet is obviously essential, and it needs to be remembered that certain foods given in excess can actually cause medical problems, for example feeding cats exclusively on liver over long periods will result in a condition called hypervitaminosis A, which leads to abnormalities of the skeleton.

Standing firm and insisting that a cat changes its dietary habits may be met with temporary indignation, but owners should be encouraged to persevere, knowing that they are doing the best for their pet. Any dietary changes should be made gradually and it often helps to start by mixing a little of the new food with the old. Gradually the ratio of new to old is increased, which not only allows the digestive system time to adjust to the new menu but also gives the cat the opportunity to get used to the new taste.

Don't expect a finicky cat to take to a new diet overnight—that will only lead you down the path of extreme frustration and disappointment.

Loss of appetite following an illness or injury is relatively common in the cat, and when nursing these individuals tender loving care is all-important as it is for cats whose anorexia is the result of a psychological rather than a physical trauma. Cats do not suffer from anorexia nervosa in the human sense. They do not perceive a need to be thin in order to be acceptable to society and they do not indulge in techniques such as secret regurgitation or self-induced vomiting in order to rid their systems of unwanted food, but there is evidence that cats will starve themselves as a result of behavioural depression due to the loss of either an owner or another animal, or because of fear associated with the arrival of a new resident (either human or feline) in their home. Where fear is identified as the cause of a decrease in appetite, treatment must include a programme of desensitisation to the fear-inducing stimulus (see Nervousness, p. 165).

Question

We recently had to have one of our beloved cats put down after a quite prolonged period of illness, and ever since our other cat has refused to eat. The two cats had been together for twelve years and had never been apart. They slept together curled up on one particular chair and they ate their meals side by side in the kitchen. Our vet can find absolutely no reason for this sudden loss of appetite and although my husband says that it must be a coincidence I am not so sure.

Answer

In the absence of any medical cause for your cat's refusal to eat I would agree with you that it is indeed connected to the recent loss of your other cat. Although some writers would lead us to believe that the cat is a totally asocial and solitary animal, we know from our own observations that this is far too simplistic a view. Those of us who live in multi-cat households are often aware of special bonds between certain feline individuals, and they certainly appear to go through a period of grieving if one of the pair dies. This is what is happening with your cat and treatment will require much patience on your part.

Sometimes replacing the other cat can help, and the survivor forms a bond with the new resident almost immediately. However, this is by no means guaranteed and for some individuals the stress of accepting a strange cat into their core territory just adds to their sense of trauma and makes things worse!

In cases of anorexia due to the loss of a companion, treatment centres around slow and steady encouragement and making the food as attractive as possible. Heating the food to body temperature will help to release appetite-stimulating aromas which can be all-important to the smell-orientated feline. The use of baby foods may be beneficial since they are easy to ingest, and adding small pieces of some highly flavoured and highly scented foods can be a useful trick. It is important to encourage the cat during feeding and to help to increase his feelings of security in the home.

In some cases the use of drugs designed to increase the appetite and decrease anxiety may be advantageous. However, as with any behavioural problem, the drug dose should ideally be a decreasing one over a relatively short period of time, in order to avoid merely masking the symptoms and leaving the cause unresolved.

Anthropomorphism

This is one of those long words with a very simple meaning that people like to drop into conversation in order to impress, and no book on behaviour would be complete without it. It means to regard an animal as having human qualities and is popularly illustrated by much loved children's characters such as Peter Rabbit and Mrs Tiggywinkle.

However, anthropomorphism is not limited to the world of make-believe, and many pet-owners grant their companions human status and expect them to be capable of human thought processes and emotions. It is often said that cat owners are prone to regard their pets as substitute children, and it is certainly true that many cats are regarded as a very important part of the family. I am sure we all agree that there is no harm in that, but when the cat ceases to be regarded as an animal, problems can result.

In many ways the fact that the cat retains such a high level of independence and keeps itself slightly aloof does help it to escape the full-blown effects of anthropomorphism and allows it to cling on to a certain amount of feline respectability. Whilst many dog owners describe themselves as Fido's mummy and daddy, cat owners tend not to label themselves with such titles since they conjure up a picture of dependence which they cannot justify.

The truth is, however, that the cat owner's relationship with her pet is essentially a maternal one: the cat takes on a kitten-like role and looks to the human as a mother figure, whereas the dog regards its people as more dominant members of a social pack. If anything, therefore, the cat owner has more cause to describe herself as the feline's mummy, but to do so is to take the relationship for granted in a way that the wild side of the cat's nature does not permit.

Question

Whenever I catch my cat doing something that he shouldn't he slinks away and then sulks for hours. He sits with his back deliberately turned and refuses to take any notice of me when I call him. I know that he can hear me since he puts his ears back, but he's just like a child who won't admit that he's in the wrong. What can I do to stop him giving me the cold shoulder?

Answer

In just these few lines you have attributed a range of human emotions to your cat and given us a picture of a sulking feline who nurses his wounded pride and refuses to admit that he is in the wrong. From your description your cat closely resembles the cartoon character Tom (from *Tom and Jerry*) who has been caught stealing food from the fridge.

There is no doubt that your cat is behaving in the way that you describe, but the clue to what is really happening lies in your reaction to the cat's misdemeanour. Having found him misbehaving you react angrily and chastise him with a harsh voice and probably a fixing stare. In feline terms staring is a most intimidating action and your cat reacts by deliberately turning away so as to avoid your staring eyes. If he were to stare back at you he would risk provoking further hostility and so he takes defensive action. Far from being a haughty, defiant individual he is demonstrating a feeling of social inferiority. By regarding your cat's posture in human terms you are in danger of misinterpreting his signals and so misunderstanding his behaviour.

Attention-seeking

There are times when the demands of our cats become somewhat irritating and in most cases demands made early in the morning or, worse still, in the middle of the night are likely to be met with little patience or understanding. Cats can be extremely ingenious in their methods for attracting our attention and by knocking over objects, jumping on owners and running uncontrollably through the house they will often achieve the desired effect. The most effective demand, however, will probably always be the persistent and penetrating miaow. Whether the aim is to make us get up and give them breakfast or merely give them a cuddle, the result is often quite a serious strain on the relationship and sometimes even on the family itself.

Question

I love my cats dearly, but the behaviour of one of them is starting to drive me to drink (not literally, you understand!). Recently I got married and my husband moved in to share the house which I have lived in with my cats for some three years now. Two of the cats accepted Richard's arrival without any problems and were pleased to accept the extra attention he provided. However Beattie, my four-year-old spayed female, seems to be having extreme difficulties in coming to terms with the new arrangements and about six weeks ago she started to howl persistently at our bedroom window from about 3 a.m. The only way to stop her is to get up, let her into the room and pet her.

My husband is fond of the cats but does not share my devotion towards them, and as you can imagine Beattie's behaviour is not contributing to a feeling of marital harmony. As if to add insult to injury, she has also taken to spraying in the house, but only on items belonging to Richard, such as his best suit and even his underwear. I really feel that I must do something before I lose either my cat or my husband and I hope that you can help.

Answer

The root cause of Beattie's problem is a combination of the disruption in her home environment and her low level of competence in dealing with the changes in her (and your) life. The spraying aspect of her behaviour arises from a need to surround herself with familiar scent and counteract the scent of your husband whom she regards very much as the intruder. This will be dealt with in detail in the section on spraying (p. 191). For the moment I want to concentrate on the howling.

The loud demands for attention that Beattie has perfected probably originated from a genuine feeling of anxiety and a need to attract your attention so that you could offer her comfort and reassurance. For some cats, the presence of their owners is all that is required, and they are comforted merely by the fact that their owners will quickly appear when called. For others such as Beattie, actual physical contact and petting are needed and they will not give up until they achieve it. It is common for such behaviour to be exhibited in the early hours since night-time is when the cat is likely to feel most vulnerable. By training you to respond to her calls, Beattie has ensured that she need not face anxieties alone and her behaviour has continued as a learned pattern because of the reward you provide by comforting her.

Treatment involves ideally ignoring and at any rate failing to respond to her demands whilst engineering an unpleasant consequence to her howls. The first objective is often easier said than done, and short of investing in some industrial-type ear defenders it is unlikely that you will be able to ignore her! Consistently failing to respond will also require unshakeable resolve on your part and constant encouragement from your partner.

The easiest component of the treatment plan is the use of an aversion technique to associate the behaviour with an unpleasant experience—options include a thin jet of water from a water pistol or a startling noise. No doubt Richard will be keen to help with this part, but both of you need to remember that the aversion must be seen by Beattie as a direct consequence of her own behaviour rather than a punishment meted out by you. You will remain neutral so that you can continue to be a positive and predictable symbol of security. Provided that the aversion is consistently delivered in a quick and quiet manner, it often takes only a few more early morning calls to get the message home. Since the attention-seeking appears to be directed specifically towards yourself, it may also help to decrease Beattie's attachment to you by only offering affection at your initiation, while at the same time encouraging a positive relationship with Richard by letting him take on the pleasant duties such as feeding and fussing.

B

Babies

The arrival of a new baby is an exciting time for any family, and it is also a time when parents have a deep desire to protect their offspring from any potential danger. These feelings are present even before the baby arrives and, especially for the mother, pregnancy brings with it an overwhelming sense of responsibility for the welfare of the unborn child. Recent publicity has drawn attention to the potential effects of a disease called toxoplasmosis on pregnant women and has also highlighted the link between this disease and the domestic cat. This has led to a dramatic increase in queries from worried cat owners about the possibility of problems resulting from owning a cat whilst being pregnant and in some cases owners have not stopped to ask questions but have simply got rid of their cats rather than take any chances.

It is true that cat faeces are a potential source of the protozoan parasite *Toxoplasma gondii* which causes this disease, but it is not the only source, and the risk of being infected by your pet cat must be kept in perspective. Other possible routes of infection include eating raw or undercooked meat, unwashed or uncooked fruit and vegetables and unpasteurised goat's milk or goat's milk products. It is therefore sensible to pay special attention to food preparation during pregnancy and to ensure that all meat is well cooked before eating it.

Most cats that hunt or are fed on raw meat are likely to become infected and although they may not develop disease symptoms themselves, they can shed significant numbers of the infective oocysts in their faeces. For this reason contact with dirty litter trays and soil that has been contaminated by cat faeces should be avoided during pregnancy. If you cannot find someone else to empty the litter tray, always wear gloves and always wash both the gloves and your hands when you have finished. Similarly, when gardening make sure that gloves are worn and that your hands are thoroughly cleaned when you come into the house. It is quite all right to handle your cat as usual, and provided normal hygiene precautions are taken, such as washing hands after handling and before eating or preparing food, you can still enjoy her company throughout your pregnancy. For anyone who has special concerns about toxoplasmosis or who simply wants to know more about it, the Toxoplasmosis Trust provides a range of literature and also offers a telephone helpline (071-713 0599). Education is vital if we are to limit the tragic consequences of the disease and at the same time prevent unnecessary concerns about cat ownership during pregnancy.

Question
Following the recent birth of my first child I have been increasingly concerned about the behaviour of my cat. He is a large three-year-old neutered tom and has a very fluffy long-haired coat. I have noticed that whenever he is in the same room as my son he sits looking up longingly at the moses basket or carrycot and I am sure that given half a chance he would jump in. What can I do to stop him and prevent a terrible accident?

Answer
For parents of newborn babies most of the fears about cats stem from an old wives' tale that they like to sleep on babies' faces because of the smell of milk and therefore suffocate them. Many people go as far as getting rid of a much loved pet because of the possibility that it might harm their newborn child, whilst others delay getting a cat as a companion because they see them as a potential threat. I do not doubt that your cat would love to curl up in the carrycot next to your son, but the reason is purely that the carrycot is a warm and secure place and not that your cat wishes your baby harm. However, if he were to get in and inadvertently smother your child's face with his fluffy coat his original motive would be somewhat irrelevant and it is of course essential to prevent this possibility.

The answer to your question is simply to ensure that your child is *never* left unattended with your cat or in a position where your cat could gain access to him—for example, through an open window or a cat flap. This rule applies not only to the newborn child, although he is perhaps seen as the most vulnerable, but also to the crawler and the enthusiastic toddler. Initially the aim is to protect the helpless child, but as time goes on it will be the cat that is the one in need of protection!

If you leave your child out in the garden in his pram, don't forget to use a cat net. Even if your cat is safely out of the way in the house, there will no doubt be other cats in the neighbourhood who might just be in search of a warm place to lay their heads. It is obviously important for your cat to learn to accept the new addition to your family, and keeping them permanently apart is neither practical nor desirable. You want the baby and the cat to learn about each other right from the start and by using controlled and, most importantly, supervised introductions, the cat should be allowed the opportunity to accept the baby's scent and the scent of all the paraphernalia that comes with him, such as the cot, the pram, the nappy bucket and so on. He needs to realise that neither the baby nor the equipment poses any sort of threat to his security or his right to expect affection from you.

For most cats, the fact that you spend a large percentage of your time attending to the newcomer will be of little consequence—after all, they are not as dependent on people as their canine counterparts and are used to receiving attention only on demand. However, for some members of the more demanding breeds, such as the Orientals, a baby may well be seen as a competitor for the owner's affections, and it is important to train the cat to accept attention on your initiation rather than on the cat's. By doing so you ensure that you are not expected always to be on tap, and by combining attention for the cat with the presence of the baby the cat will come to regard him as a friend rather than an unwelcome intruder.

Biting

Although the subject of biting usually makes one think of an angry dog hurtling towards you with an impressive set of glistening white teeth, canines are not the only creatures who can inflict quite severe injuries with their gnashers. In most cases we think of injury caused by cats in terms of deep, penetrating scratches, but a cat bite can be equally painful and can carry with it just as high a risk of infection. Cats carry offensive organisms on their teeth and claws, and it is sensible to treat any cat-inflicted injury as a potential problem and pay attention to proper cleaning of the wound.

The bacteria species *Pasteurella* is a particularly unpleasant organism, and if it becomes embedded deep into the skin as a result of a feline bite it may go on to cause septicaemia, so you can't be too careful. We saw in the section on aggression (p. 74), that cats will generally try to avoid out-and-out conflict in order to spare themselves from the risk of injury, but for some owners being bitten by their cat is a common if not daily occurrence. In many cases, not only are the attacks unexpected and unprovoked but they are also deeply distressing since they often happen when the owner is petting the cat and showing it love and affection. The cat may have been sitting quite happily with the owner for some time when it suddenly turns and bites the hand that strokes it.

Question

I own a very special little cat called Midge and for ninety per cent of the time she is the perfect pet. There is just one aspect of her behaviour which I find both intriguing and at the same time very distressing. Since I live alone my cat is my primary companion and I like to sit down in the evenings and cuddle her, but it is becoming increasingly dangerous for me to do so!

When I first pick her up and start to stroke her she purrs contentedly and lies totally relaxed in my arms, but then suddenly, and without any warning, she turns on me. Grasping hold of my arm with her front paws she kicks furiously with her back legs, scratching me in the process, and then sinks her teeth into my arm. During the attack she appears to be confused and disorientated and afterwards she jumps down from my arms and dashes away as if in a blind panic. Sometimes she stops a short distance away from me and sits down before frantically grooming herself. It is almost as if she is a Jekyll and Hyde character, and her unwarranted aggression is making me nervous. What can I do?

Answer

What you are describing in your letter is a syndrome that the Americans call 'Petting and Biting Syndrome'. Your cat is reacting quite naturally to its feelings of being trapped and vulnerable, and what we want to know is why she suddenly feels threatened by you when only moments before she was enjoying being the centre of attention. Firstly we need to recognise that this sort of behaviour can be induced in most cats; what varies is the point at which they change from accepting attention to reacting violently—in other words, their tolerance threshold. For some cats any attempt to pick them up will be greeted with a display of aggression, but for the majority the act of being handled, and indeed cuddled, at least for a short period, is welcomed.

One explanation of this sort of behaviour associates it with the individual animal's past and makes a connection between its early handling experiences and eventual tolerance threshold for human contact. Cats that have been handled from an early age and are used to being nursed usually have a high threshold and very rarely show aggression, while others, that did not have a lot of human contact in kittenhood and have been handled less frequently in adulthood, will show an aggressive reaction within a few minutes of being picked up. It has been suggested that there may even be an association between these sudden attacks and the cat's memory of an unpleasant experience. It may be that at some time in the past someone has stroked the cat gently before suddenly grabbing it and holding it down, and as a result the cat now views a friendly hand as a potential threat, and eventually the fear of being grabbed overwhelms the desire to be petted and the cat lets fly.

The other explanation for your cat's behaviour is rooted in the mutual grooming between cats. A mother cat will spend quite a lot of time licking her kittens while they are young and it is thought that the human hand is viewed by the infantile domestic cat as its mother's tongue. Comforted by the maternal grooming that we offer, our cats are content to lie back and relax in our arms, but there comes a point when they suddenly revert back to adulthood and feel trapped and vulnerable. The movement of our hand no longer resembles the calming grooming of the queen but rather the threatening paw of another cat, and so our docile pet responds with a natural defensive attack.

Treatment for your cat must be aimed at increasing her threshold of reaction and it is important to progress very slowly so as not to overwhelm her. You should start by fussing her in very short sessions without even picking her up, so that she is free to retreat as soon as she feels the need. Once she readily accepts such attention you can start to put her onto your lap and stroke her gently, but remember to concentrate on the back and the head, avoiding the more sensitive areas such as the stomach and the legs, and be sure to allow her to escape easily should she so wish. From now on you can gradually increase the length of time you spend petting your cat, but it is sensible never to restrain her forcibly, since this will only serve to lower her tolerance threshold and undo everything that you have worked so hard to achieve.

Boarding

For most cat owners, taking their pet with them when they go on holiday is not an option in the same way as it is for dog owners, although of course there are exceptions to every rule. For one thing, cats tend not to travel very well, and for another there is the chance that they will want to go off exploring and risk getting lost in the process. All in all, our cats are much happier in their own surroundings and many of us leave them at home and rely on friendly neighbours to make sure that they are fed and watered and hopefully given at least *some* love and affection in our absence. Some cat owners feel that to put their cat into a boarding cattery would be to deny it the freedom that it has come to expect, while others, worried about the risk of road traffic accidents, view a cattery as a safe and secure place where they can leave their precious companion in the sure knowledge that no harm will come to it.

Question

The opportunity has arisen for me to take a fortnight's holiday for the first time in ten years, and as you can imagine I would very much like to go. However, I am extremely concerned about what I should do with my cat, Misty, who is a very nervous individual and relies on me for constant reassurance. I recently went away for just one night and when I came home she had urinated on my duvet. I am somewhat apprehensive about what might happen if I am away for longer, and although I have some very nice neighbours who will come in and feed her, I wonder if she will be able to cope without me. On the other hand, it seems cruel to put her in a cattery where she will be confined when she is used to being a free spirit with access to the garden at all times. What do you suggest?

Answer
Take the holiday. Ten years is a long time to go without a good break and I am sure your cat would agree that you deserve it! It is only natural that you should be concerned about what to do with Misty and there are usually pros and cons to both options. Leaving a cat in familiar surroundings is often assumed to be the kindest option, but for some cats, home without their owners ceases to be a secure den because they depend so heavily on their owners as a security bridge. It would appear that Misty is one such cat and when you went away overnight she reacted to your absence by deliberately urinating on something which smelt strongly of you—in this case your bed. Other cats have been known to urinate on their owner's favourite armchair or on dirty clothes that have been left lying around. In each case the cat is engaging in a form of marking behaviour (see p. 160) and using urine to associate his or her scent with that of the absent owner. It is thought that in doing so such cats are working to present a united front to any potential enemies who may use the opportunity of the owner's absence to challenge them.

If Misty felt threatened and vulnerable when you were away for only one night, it is likely that she will be even less happy to live in the house without you for a fortnight. In such circumstances it is probably better to board her in a cattery, and you should look for one that is run to the standards suggested by the Feline Advisory Bureau. Not all catteries are maintained to such a high standard and it is always worth taking time to visit a few before deciding on which one to use. Although your local vet may have a list of local catteries they will not usually recommend one over another and therefore it is best to rely for personal recommendation on friends who have boarded their own cats.

If Misty likes to spend a lot of her time in the fresh air, you may be able to find a cattery that has individual outdoor runs available, or stimulating views to keep her amused. Most cattery owners are well aware of the need for a cat that is used to being out hunting to be adequately stimulated and to have outlets for its energy, and therefore provide a range of toys and scratching posts. Putting her into a cattery need not be an unpleasant experience and in many ways it will be the kinder option.

If you feel very strongly that boarding is not possible, or if you cannot find a suitable cattery in your area (which I very much doubt), then you could try lodging her with a friend or employing a house-sitter who will care for her, provided you are careful to make sure that any house-sitter is bona fide before leaving her in your home. If all these options fail and you are left with no alternative but to leave Misty in the house with your neighbours coming in to feed her, there are certain things that you can do to help her cope, such as practising short periods of separation during the weeks before your holiday and working to improve her ability to cope with novelty and challenge.

C

Calling

This is the term used to describe the often quite penetrating noise made by queens when they come into oestrus. Although it is perfectly natural behaviour it is none the less distressing when the novice cat owner hears it for the first time.

Question

I have recently acquired my first cat, an eight-month-old Siamese, and she has settled very well into her new home. However, recently she has started to behave in a most peculiar manner and I am worried that there may be something drastically wrong with her. I had intended to have her spayed but I am concerned that she may not be well enough to stand an operation. She starts by going off her food a couple of days before these 'attacks' and she seems to need to use her litter tray more often than usual. Her behaviour then becomes increasingly restless and she sits for long periods staring out of the window and howling. Then she crouches on the floor and, with her back end raised off the ground, treads with her front paws and screams. You can see from her face that she is in pain: her eyes become huge and her ears are flattened against her head. If I try to touch her or stroke her back she pulls her tail to one side and groans. I have not been able to pinpoint the exact source of the pain and in between these episodes she appears to be perfectly all right, but I am worried that she is suffering.

Answer

Obviously it would be sensible for you to get your cat checked over by a veterinary surgeon, but in view of her age and the apparent absence of any medical problem, what you are describing is most likely to be perfectly normal behaviour. In fact rather than postponing her operation because of her behaviour you should be getting her spayed in order to stop it!

For people who have never witnessed it before, a queen in oestrus can easily be mistaken for a cat in severe pain and queries like yours are extremely common. The 'call' of a sexually mature cat can be both very loud and very disturbing, not least because it may closely resemble the cry of a young baby. The Siamese breed, which is well known for being highly vocal at all times, does seem to have a particularly loud and persistent call and obviously your cat is no exception!

What you are witnessing in these displays is behaviour intended to attract the attention of any local toms and potential mates, and by stroking your cat along the back and especially at the base of the tail you are stimulating her to take up a receptive position, which is why she moves her tail over to one side. The posture you describe, with the back end elevated and the front paws treading rhythmically, is termed lordosis, and the accompanying vocal calls are a signal to the tom cats. Many owners report that their cats go off their food just before they come into oestrus and that they urinate more frequently and behave in a generally restless manner. Often they will be more affectionate than usual, rubbing up against objects and people and rolling enticingly on the ground. It is likely that the queen is engaging in this sort of behaviour in order to spread her scent around, since for the smell-orientated feline this is an important way of conveying the message that she is ready for her mate!

Cannibalism

The belief that tom cats will kill and eat kittens has led to the exclusion of males in breeding establishments from areas housing nursing queens and their offspring, but the truth is that such cannibalism is rarely seen with domestic cats. The theory of why infanticide occurs at all in the feline world is covered in Part Two in this book, under Infanticide (p. 50).

Question

I have been breeding cats on an amateur basis for over eight years now and have six cats that share my home. Recently the youngest queen had a litter of four apparently healthy kittens, but over a period of three days she killed and then ate each one of them. I had never experienced anything like this before, and as you can imagine I was horrified. Why should a queen destroy her own offspring in this revolting way? Is it likely that she would do the same again if I were to let her have another litter?

Answer
Deliberate cannibalism is very rare in cats and not many owners will ever witness what you have described. However, your queen has proved that it can happen and we need to take a look at the possible explanations for her behaviour. In cases where the queen herself is malnourished it is possible that her drive for self-preservation is simply stronger than her desire to rear a litter, but this is obviously not the case with a well loved pet cat such as yours. According to the Californian behaviourist Dr Benjamin Hart, it is the litters resulting from the second pregnancy in the year, and weak and sickly kittens, which are more prone to being cannibalised, but once again this theory does not apply to your situation.

A more likely explanation put forward by leading behaviourist Michael Fox is that there is a failure of the cat's hormone system to inhibit prey killing, and your queen has mistakenly viewed her kittens as prey and dealt with them accordingly. Another explanation is that cannibalism results from aggression being redirected onto the kittens after the queen has been disturbed, either by excessive outside disturbances or by another cat which frightened or threatened her in some way.

Some people believe that the stress of overcrowding can itself lead to cannibalism, even when there has been no direct challenge to the queen, and in a multi-cat household such as yours a combination of these two theories could well be the explanation. Failure of the mother to find a secure nest in which to raise her litter may also induce her to destroy them and it is therefore very important that all breeding queens are provided with a quiet kittening area away from the hubbub of the house and from any other feline residents. She should be allowed access to such a peaceful haven for at least two weeks before the kittens are born.

Although I cannot guarantee that your queen will not act in the same way to subsequent litters, it is likely that, provided she is supplied with adequate space and tranquillity, she will take good care of any future offspring.

Castration

It is now widely accepted that unless you own a cat that is to be used as a stud in a breeding environment, the sensible thing is to castrate him as he reaches sexual maturity; by far the majority of male domestic cats in this country have been neutered. Dog owners tend to be reluctant to castrate their companions for a number of reasons. These may include a fear that the operation will in some way alter his personality for the worse, a belief that castration will make the dog fat, or simply a feeling that to rid him of his virility would be unfair and even cruel.

So why is it that cat owners are only too ready and willing to take their pet to the vet for his 'little operation'? The most obvious reason is that the owners do not want their cat fathering litters of unwanted kittens. Unlike responsible dog owners who can, on the whole, control the sexual activity of their pets, cat owners recognise that they will have little if

The sensible thing is to castrate him as he reaches sexual maturity.

any control over what their tom gets up to when he's out and about. It is also likely that owners are aware of the behavioural 'benefits' of castration and want surgery performed in order to make their tom into a more 'acceptable' pet.

However, not everyone agrees that widespread neutering of the feline population is necessary or indeed justified, and those who believe that the performance of any surgery without a medical reason is wrong put neutering for convenience in the same class as declawing and label it as an unnecessary mutilation. Their argument is that the use of castration to make our cats more acceptable to us stems from a selfish human desire to enjoy feline company without accepting any aspects of their natural behaviour which we as humans consider undesirable.

Question:

My un-neutered tom cat is forever disappearing for days at a time and when he eventually returns he is always in quite a bad state of repair and often in need of veterinary attention. Although he is a very affectionate and loving cat towards humans, it seems that he has a death wish where other cats are concerned. Despite the fact that he seems to be the one who comes off worse, he continues to pick fights with any passing cat. I have even had some complaints from my neighbours about the noise that he makes and about the fact that he is often caught spraying up their back doors or worse still on their furniture when he gets in through their cat flaps.

I love him very much and have always been reluctant to have him castrated, but it seems that owning an un-neutered tom does not fit in very easily with life in suburbia! What should I do?

Answer

You obviously had your own reasons for deciding not to have your cat castrated and you have tried living with an entire tom. However, from your letter it would appear that you are now finding your relationship with your cat far from ideal. The definition of a behaviour problem is 'any behaviour that detrimentally affects the lifestyle of the owner and the lifestyle of those around them', and certainly your cat's behaviour seems to be doing just that.

Those who live with an un-neutered male cat must accept that he will come home less often, will be frequently scarred from fighting and will repeatedly develop fight wound abscesses that need veterinary attention. They must also accept that there may be components of his behaviour which could be considered less than sociable, and although these may not pose any problems for those who live in an area of low population (both human and feline) they are sometimes a source of tension in over-populated suburbia.

The simple answer to your question is, have your beloved tom castrated, and reading between the lines I am sure that you had already reached that conclusion for yourself. Basically the operation will remove the hormonal drive for specifically male behaviour and therefore make him generally more docile and less demanding. It is likely that he will become more affectionate, and many cats also show an increase in playful behaviour following castration. He will tend to stay around his core territory and spend less time roaming away from home as the size of his territory decreases. He will also lose much of the secondary hormonal motivation to fight and will be more able to cope with the proximity of his feline neighbours. His urine will become less pungent and he will be altogether more sweet-smelling, which will no doubt make him more popular with his human neighbours! As he becomes less territorial in his outlook he will also tend to spray less, but you must remember that spraying is not only a behaviour of un-neutered toms and can be a problem behaviour in its own right for some cat owners (see Spraying, p. 191).

Having outlined the many behavioural advantages of castration, let me qualify the situation with one word of caution before you take your cat to the vet expecting a miracle transformation in his character. The age at which castration is performed does have an effect on the behavioural outcome of the operation and in general the younger the cat the greater the suppression of undesirable sex-related behaviours. Castration can only remove the hormonal influence, and for a fully mature adult which is sexually experienced the behaviour is likely also to have a learned component to it. Consequently he may continue to mark territory, fight with other males and even mount females for a few months after the operation, whereas a young male castrated at or before puberty would not develop this male potential.

Chase

Chase is an essential component of the hunting process and a natural part of a cat's behavioural repertoire. As such it has already been discussed in Part Two of this book, but there are occasions when chase behaviour is seen as a problem.

Question

I realise that it is normal for cats to chase, and I have often played with my own cat, using some rolled up paper on the end of a piece of string and encouraged her to chase it. However, I am starting to get concerned about the fact that she seems to be hallucinating and madly chasing imaginary objects around the house. At first I thought she was having some sort of fit and that it was affecting her eyesight. I have taken her to the vet to have her eyes checked but he tells me there is nothing wrong. Can you offer me any explanation for these sudden outbursts which seem to stop just as abruptly as they begin?

Answer

The behaviour you are describing is not abnormal and your cat is by no means alone in exhibiting it. In fact the majority of cats that spend all or most of their time indoors will occasionally indulge in these sudden mad chasing displays. It is very common for owners to confuse this sort of behaviour with a fit, as you have done, and this is quite understandable—one minute your cat is a docile pet, the next it is an uncontrollable whirlwind. The fact that the cat suddenly stops and acts as though nothing has happened also makes it appear to be some type of seizure, and many owners take their pet to the vet convinced that there is a serious medical problem.

In fact there is a very simple behavioural explanation. What you are witnessing is something called a displacement activity. Although your cat is undoubtedly very well cared for and does not want for food, warmth, love or affection, she lacks the opportunity to express her inherited behavioural urges to hunt and indeed to flee. Spending most of her life indoors, she has no prey to stalk and catch and no enemies from whom she needs to escape. Eventually, without the necessary natural stimuli to trigger fast motion and the predatory chase, your cat reaches a point where any stimulus, however insignificant

One minute your cat is a docile pet, the next it is an uncontrollable whirlwind.

to us, opens the floodgate for her pent-up energy and she lets fly with a mad dash through the house.

Farm cats, who spend most of their time outdoors chasing real prey and interacting with rival neighbouring cats, will usually be very relaxed whenever they come indoors. They spend their time sleeping by the fire or quietly grooming themselves after their hard day's work. In comparison the urban feline spends most of her time wandering about the house and perhaps venturing out into the garden where there is usually little in the way of tempting prey. As time goes by the energy builds up and the frustration increases until suddenly the cat gets up, looks around and then chases after what seems to us some imaginary object. Having released some of her energy and exercised her hunting skills she quickly returns to her relaxed state as if nothing had happened.

Some owners tell me that these outbursts always occur at the same time of day, usually the evening, but they can identify nothing that might be acting as a trigger; others say that their cats behave in this mad cat manner after using their litter tray or after hearing a sudden sound or seeing a sudden movement. Whatever the trigger may be, and there is every chance that you will never identify it, the outcome is the same.

Treatment involves providing all indoor cats with adequate stimulation in the form of toys, activity centres and human interaction in order to allow them to release their enormous energy supplies. Your games with the string and paper will do just that.

Children

Whilst young babies need to be supervised for their own sake if there are cats around, it is likely that supervision with crawlers, toddlers and older children will be more for the sake of the cat! As the baby grows and begins to become mobile the cat gradually finds that all its traditional safe havens lose their secure feeling as the intrepid explorer finds his or her way under the table and behind the settee. Many cats will simply elevate their resting places to the top of the boiler or a high shelf, well out of the way of little fingers, but it is up to owners to ensure that there is always somewhere safe for the cat to rest.

Of course, keeping the two permanently apart is not the answer. It is important for young children to learn how to approach the cat, and how to stroke gently. Short, frequent periods of supervised interaction will help the cat to accept the right of the child to approach, and the child to understand how he should communicate with the cat.

Attention to hygiene is always important when pets are around, but the presence of children makes it a top priority. As the young child reaches the stage of putting everything in its mouth the cat's food bowl and indeed the litter tray become fascinating. Simple precautions, such as feeding the cat on top of a cupboard and covering the litter tray with a lid, can help to prevent unpleasant accidents, and of course it is vital to ensure that the cat is regularly wormed.

As the child passes from crawler to toddler stage and beyond, his movements will become increasingly co-ordinated and so less threatening in the eyes of the family cat. Parents can now expect at least some degree of compliance with their instructions and the amount of contact that the child is allowed to have with the cat can increase. He should now be starting to learn how to pick up and cuddle the cat and can even begin to take a small role in caring for her, for example by filling her food bowl each morning. By encouraging the developing friendship between the child and his furry companion, parents will be helping to educate the cat lovers of tomorrow.

Clawing

See Scratching, p. 187.

D

Den

For an animal such as the cat, which prides itself on its right to expect independence and freedom, the thought of keeping it restrained in any way may appear cruel and unjustified. The idea of an 'indoor kennel' will make most people think of the dog rather than the cat, and suggestions that the family cat should be put into one for the purpose of treating a behaviour problem may be met with fierce opposition. The reason for this is quite simply that we humans view the concept of enclosure as a punishment and equate kennels and cages with prisons. However, this need not be the case and when used properly these so-called 'cages' provide the cat with a secure den or nest in which it will feel less threatened and certainly less vulnerable.

The most important thing to remember is that the introduction of the cat to its 'indoor kennel' is vital to the success of using it in a treatment programme and sufficient time must be taken to ensure that it is carried out in an appropriate way. Simply putting the cat in and shutting the door will result in resentment and even fear, since the apparently unprovoked removal of freedom is seen as a form of punishment. Instead the aim should be to work at encouraging the cat to enter of its own free will by making the 'kennel' attractive and inviting. This is best achieved by putting familiar bedding and perhaps toys inside, together with the food and water bowls, and then feeding the cat in there without shutting the door. Once you begin to notice that he enters the 'kennel' and curls up to sleep in it of his own free will, you can start to close the door occasionally, for example while he is eating. The period of time when the door is closed can then be gradually increased until the cat is confident that the 'kennel' represents a safe haven and not a prison.

We know that the cat lives in a predominantly scent-orientated world and that in order to accept the presence of new people, animals or objects in its environment it needs to acknowledge that their scent is not a threatening one. The advantage of an 'indoor kennel' is that it enables the cat to remain in the vicinity of a novel or challenging stimulus for long enough to establish that there is no threat from it, by ensuring that he feels adequately protected. There are numerous situations that can be improved by the use of such equipment, including the introduction of a new kitten or puppy into the house (see p. 109) and the treatment of visitor-phobic cats (see pp. 107, 112, 131 and 212). Far from being cruel, the 'indoor kennel', correctly used, can be a valuable and kind way of helping cats to cope with what is often a very challenging world.

Dermatitis

Skin-related problems are unfortunately very common in both dogs and cats and there is a vast array of causes for what are often very similar symptoms. The daunting task for the veterinary surgeon is to attempt to isolate the cause rather than simply treat the symptoms, but this can be extremely difficult. Dermatology, the study of diseases of the skin, is now a highly regarded specialist branch of veterinary medicine and a number of practices throughout the country offer a referral service for these cases.

The potential causes of dermatitis range from an allergy to fleas through to complex disease processes, and a full medical investigation is essential in order to rule out any possible medical basis for the problem. In cases where such investigations have all drawn a blank, the possibility of a psychological influence must then be considered.

Question

My cat has recently been visiting the vet on a regular basis because of a problem with his skin, but despite extensive tests it has not been possible to find out what is causing his condition. My vet feels that the fact that Muffet is an extremely nervous character may be significant and he has suggested that you may be able to help.

Whenever we have visitors Muffet will hide himself away in one of our spare bedrooms and refuse to come out, and if the visitors are staying for a few days I have to take a litter tray and feeding bowl up into the room and make it into a bedsit for him! This is not a problem in itself, but I have found that while the visitors are in the house, and indeed for a few days after they have left, Muffet bites continuously at the skin around the base of his tail and pulls his fur out. By the time my visitors have gone Muffet is practically bald along his back and his skin has become very flaky and dry. I know that a simple solution would be to refuse visitors but I do not really want to do that and in any case I feel that Muffet needs help to overcome this problem rather than just avoid it.

Answer

I agree with you that simply ceasing to have visitors is not the answer, and from the description of Muffet's behaviour and the lack of any medical cause, a diagnosis of psychogenic or nervous dermatitis would seem the most likely explanation. It is perfectly normal for a cat to spend an enormous amount of time grooming himself in order to remove dirt, loose fur and indeed parasites from the coat (see Grooming, pp. 48, 129), but licking and biting at one particular location in the way that you describe is likely to be a reaction to some source of stress which the cat feels unable to deal with.

You say that Muffet is a nervous cat and he is obviously very sensitive to changes in his home environment. By spending increased amounts of time grooming himself in a very vigorous manner, which leads to irritation of the skin and significant amounts of hair loss, he is avoiding the need to face up to challenge.

Treatment for Muffet and cats like him must involve medical attention to the physical symptoms combined with a programme to increase his ability to deal with stressful situations. Initially it may help to remove him from any source of tension and if this cannot be achieved inside the home environment, a short stay in hospital may be a useful way of providing a stress-free period while treatment of the skin condition can be established.

One of the main problems in treating cases of psychogenic dermatitis is that an itch-scratch cycle develops, in which the irritation of the damaged skin causes the cat to continue scratching even after the original cause of the problem no longer exists. Using an Elizabethan Collar to prevent the cat from scratching and biting will help to break this cycle and allow the skin to recover.

Meanwhile treatment of the psychological aspects of the behaviour will involve the use of controlled exposure to challenging situations in the same way as when dealing with any nervous cat (see Nervousness, p. 165). It may help to use a decreasing dose of a suitable sedative but as always, the temptation to rely on medication should be resisted. Even when good progress is made and the cat appears to be significantly more confident about his environment, owners may still find that at times of intense stress, such as moving house, the skin problem rears its ugly head once more. However, short-term use of drugs will usually be sufficient to help the cat over these crisis points and prevent it re-establishing the behaviour.

Desensitise

This is a term that is frequently used by behaviourists. It describes the process whereby an individual is helped to come to terms with something that has previously induced a nervous or even fearful response. It involves controlled and gradual exposure to the particular stimulus concerned while rewarding the cat for its acceptance. Examples include the agoraphobic cat that is desensitised to the outdoor environment and the intensely nervous cat that needs to overcome his fear of visitors.

In each case the golden rule is to take your time: start with a very mild form of the stimulus and expose the cat to gradually increasing intensities. Desensitisation must never be rushed, and if at any time the cat responds with a fearful or anxious reaction it is important to stop at that level or even go back to a lower intensity and ensure that you have achieved acceptance before progressing any further. Any sudden escalation in the challenge that is presented may undo all of the good that has been achieved thus far.

Provided that it is carried out correctly desensitisation is a very useful tool in the treatment of feline behaviour problems and will be mentioned in the answers to many of the questions in this book.

Diet
See Chapter 2.

Dogs
Cartoonists have long benefited from the comical relationship between the cat and its traditional rival the dog, and there are countless excellent portrayals of this love-hate relationship. In most of them the dog is shown as the instigator of the trouble and cats, with tails fluffed up to twice their normal size, are seen scurrying away in search of a safe tree-top hideaway with an irate dog barking frantically as it looks longingly up into the branches.

Such is our traditional interpretation of how cats and dogs interact, but in real life situations, when the dog manages to corner the cat before it can get out of reach, sheer panic often ensues since he is not quite sure what he should do next. The cat, armed with sharp claws at the four corners of its body, can inflict quite severe injury on the confused canine and the dog will usually, and quite sensibly, back off and allow the feline to escape.

Despite the time-honoured view of our two most popular domestic species constantly at war with one another, many owners will describe how their pets live together in perfect harmony and willingly share not only the same house but the same bed. Many more describe a situation of toleration rather than actual acceptance, but the increasing numbers of families that have both feline and canine companions are a sign that the 'kill on sight' approach is not as common as the cartoonists would have us believe.

Question
I have always owned cats and would count myself as a great cat lover, but I also enjoy canine company and ever since I got married I have wanted to get a dog. Until very recently we lived in a small house with virtually no garden and we did not feel that it was a suitable environment for a dog, so we decided to content ourselves with our much loved moggie, Tip. Then six months ago we were in a position to move out of town and into a slightly larger house which has a substantial garden and we saw this as the ideal time to take on a puppy. This we did and Poppy moved in just over two months ago.

She is an absolute joy and we love her very much, but

The 'kill on sight' approach is not as common as the cartoonists would have us believe.

unfortunately Tip does not share our enthusiasm and has even gone so far as to refuse to come into the house except to eat her food, and then only if Poppy is safely out of the way. On the only two occasions when they have met Poppy has been keen to get acquainted but Tip has sat hissing furiously with her hackles up. I have had Tip for a number of years and am very upset that she feels so unhappy in her own home, but short of getting rid of Poppy (which I don't want to do) I really don't know what I can do.

Answer
First of all I have to say that I do not find Tip's reaction at all surprising. She feels that her territory has been invaded and has decided that she is better off out of harm's way. I agree with you that this is a very sad situation and all sorts of anthropomorphic feelings come bubbling to the surface! Puppies do have a tendency to be slightly over-enthusiastic in their approach to life and when you say that Poppy has been 'keen to get acquainted', I wonder just how intimidating her advances must have seemed to poor old Tip.

You tell me that Tip is still returning to the house to be fed, but if you do not see her at any other time the chances are that she has found a dog-free family who are offering her shelter and no doubt feeding her too. If you do not do something to make her home more inviting it may not be very long before she decides to move over to them completely and unfortunately, unlike our dogs, cats are more than capable of deciding for themselves where exactly they wish to live.

It is essential that you give Tip the opportunity to re-establish her rights over her core territory while learning to accept that Poppy also has the right to live with you. An indoor kennel will be a very useful tool in resolving your problem, as I outlined under Den, p. 105. It would be best to bring Tip into one room at a time, and by allowing her to remain within the protection of her 'kennel' she will be able to take time to accept Poppy's scent without running the risk of any unwanted encounters.

To start with Poppy should not be allowed anywhere near the 'kennel', but as time progresses she needs to be encouraged to approach quietly. Any boisterous or unfriendly behaviour on her part will need to be interrupted by using some form of aversion technique, so that she learns that Tip is not a toy and deserves a certain level of respect. Eventually, when Poppy has learnt to control her enthusiasm and Tip has accepted that she is not a threat, it will be possible to do away with the 'kennel', but you should always ensure that Tip has a safe haven available to her where she can hide if the going gets tough.

E

Emesis (vomiting)

Vomiting may obviously be an indication of a medical problem and anything more than a very occasional occurrence needs to be properly investigated by a veterinary surgeon. It is often far from easy to determine the cause, and when the cat spends a large proportion of its time out of doors it can be difficult to establish the underlying pattern of the problem. Sometimes, despite a good history and full medical investigation including X-rays, it proves impossible to isolate a medical root cause, and even alterations in the cat's diet or the timing of its feeds may prove unhelpful. When this happens it is time to consider the possibility of a psychological link, since psychogenic vomiting and regurgitation are well recognised phenomena.

Question

Over recent months I have been finding small piles of cat sick deposited in various rooms throughout my house, and in a household of eight cats it proved slightly difficult to find the culprit. However, after much detective work and close observation of all the cats I have discovered that Teddy, my six-year-old tabby, is the source of the problem. For some reason he has started vomiting up to three times a week and each time it seems to coincide either with a visit from our rather noisy and flamboyant neighbour or else with a family disagreement (if you know what I mean!). My vet, who is excellent with cats, has done numerous tests and has even had Teddy into the hospital to watch his behaviour, but he has been unable to come up with any solution. None of the tablets that I have been given have had any effect and attempts to change his diet failed to make any difference. In view of the timing of these episodes and the fact that Teddy is not the most confident individual I was wondering if there could be a mental problem!

Answer
It is not uncommon for cats to vomit occasionally, and for the long-haired varieties, as well as some short-haired cats, this is often a way of removing unwanted hairballs from their system. In Teddy's case the veterinary surgeon has obviously ruled out such straightforward causes and has also eliminated the range of common medical reasons. The vomiting has become quite a regular event and the fact that it occurs indoors may well be significant.

I am intrigued by the way that you describe your neighbour and would be interested to know if Teddy shows any obvious signs of distress when this particular person arrives at the house. You also mention family disagreements and your own observations would appear to suggest a link between tension in the house and Teddy's episodes of vomiting. Many humans who have been subjected to particularly stressful experiences such as taking exams, attending interviews and getting married(!) will describe how their tension and anxiety made them feel physically sick. Only in rare cases will the feeling be

It is not uncommon for cats to vomit occasionally.

carried through to the actual act of vomiting, but when the circumstances are sufficiently stressful it can happen.

Equally, for some cats life puts them under such continual pressure that their stress threshold becomes unusually low and any little tension can be enough to induce them to empty their stomach contents all over the sitting-room carpet. For some a single act of vomiting is enough to relieve the tension and allow them to get on with life, but others may continue to retch and vomit repeatedly for as long as they feel challenged.

Teddy may not show any other signs of being nervous, but you have said in your letter that he is 'not the most confident individual'. Treatment needs to aim at increasing Teddy's confidence, and where specific challenges can be identified, such as your neighbour, it will be worthwhile desensitising Teddy by employing the technique of controlled exposure (see Nervousness, p. 165). Meetings between the two should ideally take place when Teddy has an empty stomach, and an easily digested diet that moves rapidly through the system may also be helpful.

The fact that Teddy lives in a house with seven other cats may be a contributing factor since it is likely that the stress level will be higher than in a single- or even two-cat household. I doubt whether you would consider the problem big enough to rehome him, and indeed in the absence of other stress-related symptoms it would seem a very drastic first line of action, but a short-term decreasing dosage of a sedative may help to support him while the treatment is in progress. As ever, I emphasise that you should not rely on the drugs to treat the problem but rather use them as a temporary aid to the behavioural therapy.

Decreasing the tension within the home is obviously one part of the treatment programme, so encourage your neighbour to be slightly less 'flamboyant' in Teddy's company and try to minimise the 'family disagreements' for the sake of the cat. Both these steps will make a positive contribution to the situation.

Euthanasia

All pet owners dread the day when they will have to say good-bye to their beloved companion, and even when the animal is old and has had a 'good innings' its loss is keenly felt by all members of the family. I have known many people who have even resisted the desire to own a pet because they do not feel capable of coping with their eventual loss. There is a well-known saying that 'love hurts', and indeed it does, but most of us agree that the joys of feline companionship are more than worth it.

For some owners the grief of pet loss is made far more painful because they have had to make a decision to end their companion's life prematurely and taking a pet to be put to sleep is surely one of the most distressing things an owner ever has to do. Whatever the reason for euthanasia, we need to remind ourselves that feelings of grief and guilt are a natural reaction and we are by no means alone in experiencing them. However, if the family cat that has brought so much pleasure and companionship is suffering, we need to ask ourselves just who we are thinking of if we refuse to let go. When every effort has been made to alleviate its suffering and improve the quality of its life, we have done our best and we owe it to our pets not to shy away from the final decision. Humane euthanasia of a precious companion is always heartbreaking, but those owners who make the decision do so for the good of the cat, in order to preserve its dignity and say thank you for allowing them to share its life.

Excitement

At Christmas time many cat owners spend considerable amounts of money on presents for their feline friends and a variety of shops sell stockings specifically intended for the pampered pet. Along with cat food and edible treats there is usually an impressive range of toys which in some cases reflect considerable thought and imagination on the part of the manufacturers. One thing that many of these toys have in common is that they contain catnip, and indeed the inclusion of extracts from this famous plant is seen as a strong selling point. Manufacturers promise that your pet will find the toy a source of endless pleasure and excitement, and the reaction of a large number of cats seems to prove their point.

Question

I have recently bought a small toy mouse for my cat Ringo, and ever since I gave it to him he seems to be permanently excited. I have bought countless toys for him in the past but never before has he reacted like this. My friend tells me that it is because of the catnip which is in this particular toy and she has suggested that I should grow some in my garden for him. Is it true that catnip can make cats this excited and if so, why do they have such a strong reaction to it?

Answer

It sounds as if Ringo is having the time of his life with this toy and that you have made a good buy! It is true that the excitement he has been showing is a direct result of the catnip inside the little mouse and his reaction is totally natural. In common with a large number of other cats Ringo is responding to an oil called nepetalactone contained within the catnip plant and it is this that is impregnated into his toy.

This unsaturated lactone is thought to act in much the same way as some of the excitement-inducing drugs used by certain members of our own society, although the exact mechanism of its action on the feline brain is as yet unknown. Certainly people have described their cats as entering a period of ecstasy and have likened the reaction to a 'trip', but it has to be said that such comments do represent a rather anthropomorphic viewpoint! We know that a similar reaction can be seen in other feline species such as lions, jaguars, pumas and leopards, but it does not occur with tigers or bobcats.

Not all individual cats respond in the same way and you may find that your friends' cats consider the mouse less inviting altogether. The tendency to react is genetically controlled and it is thought that 50 per cent or more of the feline population show the response. The intensity of the reaction will vary from cat to cat, tom cats generally appearing to become more excited than females, but the behaviour is seen in both sexes and in both neutered and entire individuals. At one time the reaction was believed to be related to sexual responses and catnip became known as a feline aphrodisiac. Certainly some aspects of the reaction, such as the rolling, treading and flehmening are reminiscent of the female in oestrus, but it is unrelated to sexual behaviour.

Catnip (*Nepeta cataria*), or catmint as it is also called, is a member of the mint family and grows as a weed throughout the temperate zones of Europe and North America. It is easily cultivated and is commonly grown in gardens and even as a houseplant. Young kittens are said actively to avoid the plant, but over the age of about three months a certain percentage of cats will go wild every time they encounter it, while those that are not so affected will usually just ignore it.

Cats that do react usually do so immediately they smell the plant and the reaction can last for between five and fifteen minutes. The cats will first sniff at the catnip and then lick, bite and chew it in an increasingly frantic manner. Next they will rub up against the plant or impregnated toy with their cheeks and then with their whole bodies, at the same time purring, growling and miaowing loudly. Many cats will roll over and some have even been reported to leap up in the air in excitement. Some enter a trance-like state and may even sit and stare into space, while others will chase imaginary prey. It is believed that these cats are in a true psychedelic state, but unlike the drugs used by humans the effects of catnip are short-lived, quite harmless and also non-addictive.

F

Falling

The agility of the cat is legendary and one can only admire his ability to walk along a cluttered mantelpiece and not so much as move the ornaments. Whereas the over-enthusiastic tail-wagging dog plays havoc with low-lying nicknacks, the cat can live in relative harmony with household trinkets. When our cats are out and about in the garden many of us spend hours marvelling at how they can walk along the narrow tops of fence panels and jump so effortlessly from the top of the garden shed to the garage roof. All these things add to our fascination for this wonderful species, but perhaps the aspect of feline agility that we find most intriguing is the cat's ability always to fall on its feet.

Question

I hope you don't mind me writing to you since my request is more for an explanation of cat behaviour than for treatment of a behaviour problem. My cat Lilly is a very adventurous little soul and she does insist on sitting on the sill of my open bedroom window. Of course I try to stop her but I can't be watching her all the time and I am very frightened that she may lose her balance. I have often heard people remark on the way that cats always land on their feet and I was wondering if that would prevent her from doing herself any harm if she did fall from the window. I would also be fascinated to know why cats seem to be able to land the right way up!

Answer

There can be no guarantee that Lilly would survive a fall from your bedroom without any ill effects. Over the years I have treated a number of cats for injuries sustained after leaping

Cats seem to be able to land on their feet, no matter how far they fall.

out of first floor windows for whatever reason, but the remarkable thing in many cases was that they were alive at all. Certainly the ability of cats to right themselves before they hit the ground enables them to survive falls of often quite remarkable distances, but the injuries incurred can be serious.

The intriguing righting ability of the cat has been recognised for many years and at one time was thought to be a symbol of feline supernatural powers. In the town of Ypres in Belgium part of their annual festival in celebration of the cat involves a centuries-old practice of throwing cats from the top of the town's Cloth Hall tower which is some 230 feet high. It is thought that the tradition originated in the year 962 when cats where thrown from a tower as a symbol that the practice of cat worship had come to an end. Thankfully today the cats that are used are only toy ones, but up until 1817 live animals were thrown as part of this bizarre ritual.

The fact that the cat can land on its feet after a fall is due to a special 'righting reflex' which is thought to be an adaptation for survival in a species that often chases its prey along precarious branches. Inevitably there may come a time when the cat loses its balance and falls from the tree and the righting reflex ensures that such falls are not fatal. As he starts to fall the cat's body undergoes an automatic twisting movement. Firstly the front half of the body rotates and brings the head into an upright position and then, after bending up the back legs, the hindquarters twist round in line with the front. Hence all four feet are now ready for landing and just before the cat makes contact with the ground it stretches out its legs and arches its back, thereby reducing the force of the landing impact. It takes only a fraction of a second for all of this to happen and many impressive photographic studies have been used to identify the separate stages of this remarkable reflex.

So you can see that Lilly is equipped with a fantastic ability to deal with an unexpected fall, but I still feel that the best thing in your case would be to stop her from using this particular resting place.

Fear aggression

The word 'aggression' conjures up an image of a tough guy picking a fight with some unsuspecting passer-by, but aggression need not be offensive in nature and may result from fear and be used as a defensive action. In general fear will result in one of three reactions—flight, fight or freeze—and the first of these will be discussed further under Nervousness (p. 165).

Flight is the reaction of choice in most situations, but when the escape route is blocked and the cat finds itself confronting a fearful stimulus, it will either crouch down and remain motionless or lash out with its front paws, hissing and spitting as it does so. This last reaction is commonly termed fear aggression and it is a common behavioural expression in the domestic cat. It can be distinguished from other forms of aggression not only by the circumstances in which it occurs but also by the defensive posture of the attacking cat. This individual is using aggression to fend off a frightening stimulus, and having startled the 'enemy' with his display the cat will try to seize an opportunity to escape.

Question

Freddie is normally quite a gentle cat albeit slightly withdrawn at times, but whenever I try to put him into his cat basket to take him to the vet or the cattery he suddenly becomes like a wild thing and scratches my hand until he draws blood. If he gets the chance to escape then he bolts like a frightened rabbit and rushes out of the cat flap and away. I can't tell you the number of times I have had to cancel appointments with the vet because he has gone missing! I keep the basket hidden away when it's not being used, but he must have a very good memory because as soon as he sees it he tries to make a mad dash for the door.

Last week his behaviour nearly caused a major disaster as I was booked to go on holiday and, running late as usual, was taking him to the cattery at the last minute. As I tried to put him into his basket he attacked my hand and managed to escape. Thankfully I managed to find him in the garden and catch him quite easily, but he very nearly made me miss my flight! Is there anything I can do to improve his reaction to the basket before my next holiday?

Answer

The problem with cat baskets is that for the majority of cats they only appear when a visit to one of two places is about to take place. Either your owner is going to leave you with some stranger and disappear for days on end or else you are going to be manhandled on a consulting room table and more than likely subjected to one or more injections. Neither of these experiences is particularly pleasant for the cat and many will rapidly build up an association between the basket and the subsequent ordeal.

In order to treat this problem it is necessary to form pleasant associations with the carrying basket and if possible with the destinations. It is worth looking critically at the type of basket that you are using since not all varieties suit all cats. Some prefer to be able to see out and take notice of what is going on around them, while others prefer to be fully enclosed and protected. You need to match the basket to the individual cat.

Once you have done that, the basket should be left out in the house where Freddie will be able to see it regularly without anyone even attempting to put him into it. Gradually he will begin to accept its presence and may even begin to rub up against it in order to imprint his scent and make it part of his surroundings. Once you have reached this state of acceptance you should start to associate the basket with pleasurable activities such as eating, and by placing tasty morsels into the open basket and leaving Freddie to explore in his own time he will soon be happy to approach it and even take food from it. Gradually you can reach the situation where Freddie is actually fed in the basket, so making it a very desirable place to be!

Having overcome the fear of the basket itself, you need to deal with the fear of the vet or cattery which led to the problem in the beginning. Usually a visit to either of these places will involve a ride in the car, which is in itself a frightening experience for most cats since they are rarely taken out in the car for any other reason. Owners of puppies make a point of introducing their pets to the car from an early age since they want them to be happy to travel in it, but few owners of kittens follow the same procedure. Although you may not need to take your cat out on daily excursions or to the park for a walk, there will be times when car travel is necessary and early exposure will make it far less traumatic for all concerned.

Forming pleasant associations with the veterinary surgery and the cattery will involve co-operation from the staff and most will be only too happy to assist. As far as the veterinary surgery is concerned it will help to request an appointment at the end of surgery when you are likely to be less rushed, so that Freddie can be allowed time to explore the consulting room before any treatment is given. In general the less restraint the better, but obviously it is essential to pay attention to the safety of both the owner and the vet and prevent unnecessary injuries!

Offering the cat food while he is on the examination table can help to overcome fear, and if necessary low doses of sedative given before the visit can increase his tolerance so that he is better able to learn how to cope with the challenges of the consulting room. Such sedative treatment should be a temporary measure and should not be used as an alternative to behavioural treatment.

In cases of particularly fearful cats I have encouraged the owners to bring their cats to see me outside normal consultation hours so that they do not have to deal with a bustling waiting-room, and also to bring them occasionally when they do not need medical treatment so that not every trip to the surgery results in a painful injection or undignified examination.

Fighting

We saw in the section on aggression (p. 74) that in general cats will go to great lengths to employ non-violent communication in order to prevent out-and-out confrontation in any dispute. However, anyone who has been woken at two o'clock in the morning to the sound of a local cat fight will know that sometimes diplomacy fails and the fur starts to fly. In any veterinary practice there are certain cats that come back time and time again with nasty abscesses as a result of a fracas with a neighbouring cat and many owners will say that they know who the aggressor is! In fact it is common for housing estates to have a resident despot who terrorises the local feline population and who makes neighbourly relations for its owner somewhat tense on occasions.

We acknowledge that our cats are keen to defend their territory and accept that it is unlikely they will form lasting friendships with the cat next door (although such strong pair bondings can occur), but we hope that they will be able to live in relative peace. Most cats in highly populated areas manage to operate a successful time-share arrangement whereby a number of cats share common paths through their territories but only utilise them at specific times. Scent marks are used to convey messages between the cats as to who was last in the area and at what time, and by reading these signals the cats are able to ensure that they can all move about their patch without needing to confront one another. Certain more dominant individuals will have the right to patrol the area at key times such as dawn and dusk when potential prey is more likely to be found, and other cats will respect this right in return for free access through the territory at other less important times. In this way our pets manage to maintain a relatively harmonious existence and all is well until someone like Jake moves in.

Question

I am writing to you in a state of desperation and hope that you will be able to offer a solution to my problem before I am hounded out of my village by irate neighbours or made bankrupt by their demands that I pay their vets' bills! My

If he were a human he would be in Parkhurst by now . . .

problem comes in the shape of a five-year-old neutered tom cat called Jake, and although I love him very much I am rapidly coming to the conclusion that he will have to go.

The reason for this pessimistic outlook is that Jake has become a regular little yob. Not content with beating up the local felines in their own gardens, he has now started breaking into houses through cat flaps and open windows and setting upon innocent victims in the comfort of their own homes. As if such violence were not enough, he then proceeds to spray all over the houses that he has entered and leave his unmistakable calling card! If he were human I'm sure he would be in Parkhurst by now, and from some of the comments from my neighbours it would seem that they would all be in favour of him receiving a life sentence. Is there any hope of me reforming my little terrorist?

Answer

I can assure you that Jake is not the only yob in the cat world and that you are not the only owner who suffers from strained relationships with neighbours on account of their pet, but no doubt such reassurance is cold comfort to you just at the moment. It is likely that the aggressive attacks and accompanying spraying that you describe have become learned patterns of behaviour following the success of his break-ins to date!

Treatment of individuals like Jake can be surprisingly successful but it necessitates high levels of co-operation from all the local cat owners. If you take the trouble to explain to them that the treatment is aimed at reforming Jake into a socially acceptable feline they will no doubt be more than happy to assist.

You must closely control the times when Jake is allowed access to his outdoor territory and make sure that they coincide with times when the local cats are inside. By liaising with other owners you can ensure that they lock all cat flaps and keep windows closed during the times when Jake is out and about, so that the other cats can feel safe in their own homes.

It is sensible to deny Jake access to outdoors at times when his territorial instincts are likely to be most heightened, for example at dawn and at dusk, and by adjusting your cat flap you will be able to ensure that once he returns home he will not be able to leave again until you are there to open the door.

It will also help if you can discourage your non-cat owning neighbours from allowing Jake into their homes, and indeed request that they actively reject him by using water if necessary. Any attempts he makes to enter the gardens of neighbouring feline residents should also result in an unexpected shower, and owners should be asked to feed their cats in the security of their own home and not leave any food outside where it might attract Jake. At the same time you should increase his perception of your house as his core territory by feeding short and frequent meals throughout the day, so reducing his need to try and compete with other cats for their food, and also increasing his bond with you.

As I have suggested in chapter 2, there are some circumstances when the cat's diet can be suspected of affecting its behaviour patterns, and lowered thresholds for aggressive reactions are one such situation. For this reason it may be worth changing Jake's diet to one of fresh chicken and fish, and avoiding canned food which may contain preservatives or other artificial additives that are capable of causing an allergic reaction.

During the day, when the other local cats are out and about, Jake should be put outside in a pen so that his efforts to start a fight meet with failure and the other cats can approach in the sure knowledge that he can do them no harm. In this way they can gradually establish their right of occupancy in the shared territory and Jake can learn to accept their presence.

By obtaining the co-operation of your neighbours in applying this treatment plan you will hopefully begin to reform Jake and also improve relationships with the people around you, so that Jake can stay with you and you need not look for another house!

Following

Question

I have always laughed at the idea of putting a collar and lead on a cat and taking it for a walk, and even ridiculed those people who did so for being slightly on the eccentric side. However, some fifteen months ago I acquired a new kitten and she has taken to coming with us when we walk our dog. This is not something that I have enticed her to do but something that has started on her initiation. After all the times that I have chuckled at the sight of a cat being taken for a walk I now find that I am becoming a regular Pied Piper. Is there some behavioural explanation for Kitty following us in this way, or was I wrong to believe that cats do not want to be exercised by their owners?

Answer

The sight of cats being taken for a walk by their owners is not as uncommon as one might imagine and many pet shops do a roaring trade in cat harnesses, collars and leads. It has even been known for some books on cat ownership to give details on how to train your kitten to enjoy its daily exercise. Most owners like yourself feel that such treatment is better suited to the dog, and indeed it is. The cat is not a pack animal and travelling in groups does not come naturally in the same way as it does for dogs. The adult cat on the move is basically a solitary creature and we cannot expect it to want to come with the family on long communal walks.

Having said that, there are certain cats, like Kitty, who take it upon themselves to join in and who will follow quite happily in their owners' footsteps. The most likely explanation for these outings can be found by looking at the way newly mobile kittens are introduced by their mothers to the area immediately surrounding the nest. Before venturing out alone into the big wide world, kittens will follow the queen on short outings away from the nest, and by keeping a close eye on them she makes sure that they do not stray too far from home. Kitty may be approaching adulthood in terms of her age but in behavioural terms she will always remain very much a kitten (see Juvenile behaviour, p. 149) and consequently she looks on you as a surrogate queen. When you set off with the dog for your daily walk Kitty is reverting back to kittenhood and following you out into the wider territory.

In most cases cats which behave in this way will only follow for so long before they feel that they are straying too far from home and will return to the safety of their nest.

G

Greeting

There is nothing that quite compares with the unconditional acceptance of your pet cat at the end of a particularly fraught day at the office. When all around you seem to be going steadily mad the warm feline greeting helps you to put the worries of the day behind you and begin to relax.

Many cat owners marvel at the way in which their usually aloof and independent pet can be so inviting and welcoming, and I get various questions about why cats greet us in the way they do. This is not a behavioural problem but judging by the number of inquiries I receive, it is obviously an area of feline behaviour that fascinates us and a common reason for asking, why does my cat?

When owners come home to find the cat asleep on a chair in the sitting-room the feline greeting will probably consist of rolling onto its back, yawning and stretching and then staring up at the owner while slowly twitching its tail tip. This greeting is one that is reserved for people the cat knows well and it is a very friendly and trusting signal. In feline terms, exposing the belly is a sign of great vulnerability and for the sleepy cat to roll over and invite contact with this important part of the body is to show that it is willing to take a risk. The yawning and stretching are part of the waking up process and the twitching of the tail is thought to indicate a degree of conflict within the cat which may suggest that he is not completely at ease after all!

Although this inviting posture may make one feel tempted to tickle the furry outstretched stomach, many owners recognise that to do so would be to ask for a swipe from an angry paw. It may be one thing to indicate trust by exposing the belly but it is quite another to allow actual physical contact and many cats feel that even their owners are not worthy of such a privilege.

If, instead of being asleep when its owner returns, the cat is wide awake and active, then the greeting will take a different form. The cat repeatedly rubs its body up against its owner's and brushes the side of its face along the owner's legs before entwining its tail around them. As the owner reaches down to stroke it, the cat will increase the rubbing and often nudge forcefully with the corners of its mouth before wandering off and sitting down to wash itself thoroughly.

When greeting is taking place between two felines, the contact will be face to face and the fact that our cats rub around our legs is merely an adaptation resulting from necessity since we are too tall to allow them to rub our faces. Small kittens who are trying to greet their mothers meet with similar difficulties, although the height differential is substantially smaller. In order to overcome the problem the kittens will make a deliberate hopping movement, lifting their front legs off the ground as they raise their heads to make contact with their mother's face, and she in return assists them by lowering her head.

The same movement is often seen in cats that are trying to greet their owners, and many will go a step further and jump up onto the backs of chairs in order to make contact with their owner's face.

All this friendly display makes owners feel very special, but what we often do not realise is that by rubbing up against us in this affectionate way the cat is actually exchanging scent with us. For a smell-orientated feline it is important that all members of the household have a familiar scent and so he marks us with his own scent by using the special glands situated on the temples, at the base of the tail and at the corners of the mouth. Once he has done this he must then read our olfactory signals, and he does this by carefully licking the flank fur which he has just rubbed up against us and thereby 'tasting' our scent. So we can see that the greeting behaviour of our pet is an intricate mixture of vocalisation, body language and scent communication, designed as much to make the cats feel secure as to make us feel wanted.

Grooming

Unlike the dog who makes enormous demands on his owners in terms of brushing, bathing and generally keeping tidy, the cat sees to her own needs, and those of us who have ever owned a white cat will have frequently marvelled at how it manages to keep itself so clean.

For each individual cat the sequence in which it cleans all the parts of its body is usually quite predictable and it is fascinating to watch a cat in the process of grooming. The amount of time spent will obviously vary from cat to cat, but there is no doubt that this activity is of considerable importance. Indeed, it is one of the first skills that a kitten learns and from as early as three weeks of age it will begin to groom not only itself but also its littermates and mother. The act of mutual grooming helps to establish bonds between mother and offspring and this behaviour is carried on into adulthood when it is used to strengthen relationships between cats which have to live in close proximity to one another. No doubt it makes sense to get someone else to clean those inaccessible parts of the anatomy such as the back of the ears, but there is far more to it than that. Grooming is one of the most important social interactions and enables cats to communicate with each other by sharing their scents.

The tongue and front paws are the primary tools used in caring for the coat while the teeth and claws also come into play if there is particularly stubborn debris embedded in the coat. Cleaning of the head region is achieved by licking the paw and forearm and then rubbing them over the face and ears, and cats use their remarkable flexibility to enable them to reach almost every other part of their body.

As well as removing any loose or broken hairs, the act of licking is believed to stimulate the growth of new hairs from the follicles in the skin, and any cat that stops grooming for whatever reason will soon have a very poor and dull-looking coat. Failure to groom is often a sign of a medical problem and cats that rarely groom or do not do so effectively are far more likely to suffer from parasite infestations since removal of fleas and lice from the coat is one of the functions of grooming.

Grooming also has a part to play in temperature regulation for the cat, which explains why they tend to groom more in warm weather or after periods of extra excitement or activity. Whereas we humans lose heat via sweat from the many glands all over our bodies, cats rely on saliva for their evaporative heat loss. By licking the fur they deposit saliva on their coat which can then evaporate in much the same way as sweat does. When cats are sitting in direct sunlight grooming is increased, not only because of this heat regulation but also as a means of ingesting small quantities of vitamin D which is an essential nutrient. The vitamin is synthesised by the action of sunlight and by licking the coat the cat picks up this dietary additive on its tongue.

As well as licking, cats will often pull quite forcefully at their coats during grooming, and besides helping to dislodge particles held in the fur this activity actually stimulates special glands situated in the skin at the base of each individual hair. These glands produce secretions which give the fur its waterproof qualities and therefore by grooming the cat is protecting itself from the meteorological extremes of strong sunlight and pouring rain.

As we saw in chapter 5, the purposes of feline grooming are many and varied and it is not surprising that our cats spend so much of their time on this vital activity, but for some individuals grooming seems to become an obsession and problems can occur.

Question

I have noticed for a long time that whenever I have been cuddling my cat he jumps down and immediately starts to groom himself. He is a long-haired cat and I have always attributed his preoccupation with grooming to the fact that there is always a lot to be done with such a coat. About three months ago I obtained another cat and although they do not seem to be overtly fond of each other there have not been any fights as such. However, at about the same time as Fred came to live with us, George, my original cat, increased his grooming almost to the point of obsession and now whenever I have visitors George will groom himself continuously. To start with this was something of a joke and my friends commented on how he wanted to look his best when I had company, but now I am beginning to wonder if his behaviour is normal.

Answer

You comment on the fact that George has always groomed himself after he has been cuddled by you. This is perfectly natural feline behaviour. Any cat, irrespective of its coat type, will respond in this way if it has been handled by its owner, since the grooming is not designed so much to rearrange its ruffled fur as to exchange vital scent information. We live in a world predominated by visual communication but the cat relies heavily on olfactory messages and when we stroke our pets we deposit some of our own smell onto their coat. By grooming himself George is reading all the information contained in your scent and also re-establishing his own scent onto his fur. In your letter you appear to be connecting this behaviour with what you now consider to be the problem of George grooming himself excessively when you have visitors, but I would suggest that the motivation is quite different.

Frequency of grooming is known to increase when a cat becomes agitated or frightened, and in these circumstances it is described as a displacement activity. The grooming is not being carried out for the benefit of the cat's coat but rather as a means of relaxing the cat so that it can cope with some perceived conflict or tension. The fact that George started to groom more when Fred moved in is significant and although you have not witnessed any out-and-out aggression between them, that is not to say that there have not been any confrontations. In any event it is likely that there is at least an air of tension in the house and George is responding by grooming himself.

Treatment must therefore be aimed at encouraging him to accept Fred and realise that he is not a threat, and this is best achieved by using controlled exposure techniques. Using an indoor kennel to provide George with a safe and secure base may be helpful, so that he can feel protected while he is given time to come to terms with his new 'friend'.

The next step is to help George cope with the invasion of his territory by visitors, using controlled exposure and gradual desensitisation, and this will no doubt be easier once the tension in the house has been decreased. The fact there is a clearly identifiable source of anxiety in George's case makes it likely that treatment will be successful, but he may still be prone to over-groom in future if he finds himself in a state of conflict.

One potential side-effect of excessive grooming in any cat is the formation of hairballs in the digestive system and obviously the risk is increased in a long-haired individual such as George. It is perfectly natural for hairballs to form and the majority of cats will be able to vomit them up with no difficulty, but when a cat engages in more than a normal level of grooming there is a risk that the hairballs will become unusually large and form an obstruction. In extreme cases it has been necessary to perform surgery in order to dislodge them, and I would therefore advise you to take the time to brush and comb George regularly so that excess fur is removed and the hairball risk is minimised.

H

Homing instinct

They say that moving house is one of the most stressful experiences for the human race and it can be equally true for our feline friends. Many of the phone calls that I receive are from worried cat owners who want advice on how to make the forthcoming house move as painless as possible for their pet. It is always sensible to arrange for the cat to stay either in a local cattery or with friends on the day of the move so that you have one less thing to worry about and the cat is spared the trauma of moving day. If the cat is at all nervous in character it is worth considering extending the stay, so that he goes away before the packing starts at the old house and does not return until you are settled in at the new one. Ideally everything should be unpacked and organised before the cat is introduced to its new environment, but of course this does presuppose that you can unpack from a house move in less than a month or two!

Once the cat moves in it is sensible to confine it to the house for a week or so in order for it to become accustomed to the layout of the house and begin to imprint its scent on its new core territory. Eventually the time will come to let the cat out in order to investigate the wider territory and establish its right to share the patch with the residents. This is best done when the cat is hungry since it will be less likely to wander far from its home base when there is the prospect of being fed.

Question

Just over a month ago we moved house but stayed in the same town and our new house is only a couple of miles away from the old one. We are very pleased with our move, but our cat

Rainbow seems to be less than impressed. Even after all this time he is still going back to the old house and my husband is getting tired of going to collect him when the new occupants phone us at eleven o'clock at night to tell us he's there. Is there anything we can do to persuade Rainbow that this is his home now, or should we ask the people at our old house if they would like to take him on?

Answer
From Rainbow's point of view the problem with moving only a short distance is that he is likely to come across his old pathways while exploring his new territory. When he does so he will simply follow those paths and arrive back at your old house, surprised to find that things have changed and you are no longer there. If there is a cat flap at the old house, Rainbow will let himself in in a very confident manner and be quite content to doze off in familiar surroundings.

Often the problem of cats returning to their previous homes is made worse by the new occupants who inadvertently encourage them by providing food and affection.

The fact that you have considered asking the new owners if they would like to take Rainbow on suggests that they are already showing signs of being fond of him. However, if you really want to keep him with you then you must ask them not only to stop feeding him but also actually to discourage him by being less than pleasant, throwing him out and if necessary even throwing water at him. It will also help if your previous neighbours behave in a similarly hostile way towards him and refrain from greeting him when he turns up.

Keep Rainbow housebound at your new home for an extended period of about a month and during that time supply him with plenty of attention and short, frequent meals in order to strengthen his bond with you and all your family. When you do let him out make sure that he has been starved for the previous twelve hours and is therefore suitably hungry, and do not allow him to go out more than once a day. Restrict his outdoor expeditions to half an hour before calling him back and immediately feeding him. The aim of all of this is to increase the perception of your new house as Rainbow's core territory where food and affection are freely available, and so contrast it with the old house where the reception is uninviting and even hostile.

For some cats, the call of the old territory is so strong that even after a month they will still find their way back. Boarding such cats at a cattery which is a long way away from either home has been suggested as a way of attempting to scramble the homing mechanism. Other suggestions are to take a very indirect route when bringing the cat back to the new house, but there is no guarantee that such tricks will work.

The main thing is for you to stay calm and patient, since it may take a few more weeks or even months before Rainbow is as settled as you are. If, after a lot of effort on your part, he still prefers to go back to the old haunts, it may be more sensible to allow him to move in with the other family, but at least you will know that you did everything possible to keep him. We need to remember that at the end of the day we do not own our cats in the same way as our dogs and they are more than capable of making their own decisions. Maybe next time you should consider moving to the other end of the country rather than just down the road!

Homeopathy

Alternative medicine has attracted much interest over recent years, not only in the medical field but also in veterinary circles. There is an increasing number of veterinary surgeons who are members of the British Association of Homeopathic Veterinary Surgeons and specialise in the use of alternative treatments.

Drug therapy is known to have a potential place in the treatment of behaviour problems provided that it is used in conjunction with and not to the exclusion of behaviour modification techniques. One area in which they have been found to be helpful is treatment of nervous cats where short-term, gradually reducing doses of sedatives can be used to aid the learning process by removing the initial fear and allowing the cat to get maximum benefit from controlled exposure techniques. Obviously the danger is that the cat's new-found competence could become dependent on the sedative and this is a situation that must be avoided. For this reason there has been increasing interest amongst behaviourists in the use of alternatives to potentially addictive sedatives and homoeopathic treatments and Bach Flower Remedies have given encouraging results. Even if the scientist within finds it hard to comprehend just how alternative therapies work, the fact is that they do, and provided advice is taken from a suitably experienced veterinary surgeon, in the same way as when using conventional medicine, they can be a very useful adjunct to treatment of behavioural problems.

House-training

This is one area where kittens win hands down over puppies since most of them arrive at their new home at the age of six to eight weeks already fully house-trained and scrupulously clean. This is probably the single most important factor in encouraging cat ownership as an alternative to owning the traditional number one companion, the dog.

The vast majority of cats will never make a mistake, and when they have access to outdoors and use the garden for their toileting needs the owner need never face the prospect of cleaning up after them. Even when the cat is kept totally indoors it will use a litter tray and carefully bury its excreta so that human contact with the by-products of pet ownership can still be minimal.

In order to understand why it is that kittens come to us already house-trained we need to look at their strong inbuilt desire not to soil the nest. It makes good sense that they should not soil the area in which they rest since not only is the prospect of lying in close proximity to waste unpleasant, but it may also put them at risk of infection and therefore decrease their chances of survival. The very young kitten is physically unable to urinate or defaecate without the stimulation of its mother's tongue licking the abdomen and perineal region. This so-called urogenital reflex ensures that the kitten only eliminates when the queen is on hand to clean up the mess, so keeping the nest free from dirt and avoiding a build-up of odours that might attract predators.

This maternal function may continue until the kittens are five weeks old, but as they become more active and start to move about, their own activity provides sufficient stimulation to elicit excretion and most kittens are able to urinate and defaecate voluntarily by the time they are three weeks old. When the young offspring begin to wander away from the nest the queen will continue to stimulate their bodily functions but will only do so outside the nest from now on. Indeed the mother may even carry the kittens out of the nest specifically to make them eliminate and in this way she teaches them that to soil the nest is wrong.

The next stage in feline house-training is for the kittens to learn how to dig holes that can be used as latrines and how to cover them up after use. All young cats have a natural desire to rake at material which is loose and soft in texture, and it is by observing their mother's toileting behaviour that they develop this raking motion in order to excavate holes in suitable latrine locations. The smell of the excreta of others attracts them to suitable places such as provided litter trays and kittens rapidly progress in their toilet training.

Once the kitten moves on to its new home the association with litter usually transfers easily to the soil in the garden and the adult cat learns that the whole house is to be regarded as his 'nest' and must therefore not be polluted. If the cat is to lead a totally indoor life it will simply utilise the trays provided and keep the rest of the den immaculately clean.

This is the theory of feline cleanliness, and by far the majority of cats live up to their reputation. In fact it is precisely because of the cat's well-known fastidious attention to hygiene that cat owners find breakdowns in toilet training so acutely upsetting. The dog owner automatically blames himself for any mishaps that his dear dog may make on the carpet, since it is his responsibility to ensure that the dog has adequate access to outdoors. Most cat owners will also accept the odd accident, especially if it is related to a period of illness or the cat being inadvertently shut in, but when lack of house-training becomes a more persistent problem then the cat-owner relationship begins to suffer.

Question

Last year my husband bought me a beautiful Persian kitten for my birthday. I have never owned a cat before but I had always thought that unlike puppies they came ready house-trained. Unfortunately this did not appear to be the case with Gizmo who from the very first day started to go to the toilet throughout the house. At first I thought he was upset at being separated from his mother and I made excuses for him and quietly cleaned up the mess. However, I am now beginning to lose my patience since he is nearly a year old and the 'accidents' are still happening regularly. I have provided him with a litter tray in each room of the house in case he is just not able to get there in time, but it doesn't seem to help. I waited a long time to get a cat and although I am very fond of Gizmo I almost feel that I have been cheated. Cats are supposed to be clean and easy to care for, but in addition to grooming his impressive coat I am having to clear up after Gizmo more as if he were a dog than a cat. Why does he refuse to use his litter tray like any other cat and is there anything that I can do to reform him?

Answer
Lack of house-training is one of the most common behaviour problems experienced by cat owners, so rest assured Gizmo is not alone. In his case it would appear that he never learned the basics of associating litter trays with toileting, and in fact the Persian breed does seem to be somewhat over-represented in problems of this sort. Quite why this is so remains unclear, but it has been suggested that there may be an inherited inability in this breed to investigate loose litter at an early age. Alternatively the kittens may simply be unaware of the fact that they should be imitating their mother's toileting behaviour, or if the mother herself has a poorly developed association with litter she may be failing to set an example to her kittens. Without the opportunity to learn by observation such kittens will grow up with an equally poorly developed association and so it will go on for generation after generation.

This last explanation is a very plausible one, but it cannot be the whole story since there have been cases where perfectly house-trained Persian queens have produced one or more kittens that have had house-training problems. Whatever the reason, the fact remains that some kittens, Persian or otherwise, fail to become litter-trained before they leave the breeder and as a result pose quite severe problems for their subsequent owners.

In order to treat kittens such as Gizmo you need to go right back to basics and use the feline instinct of not soiling the nest as the foundation for a treatment programme. To begin with Gizmo should be confined to a small pen in which there is only a bed and a small area of floor that is covered with a suitable litter. At mealtimes he can be brought out of the pen to feed and be given love and attention by the family. When inside the pen he will have the choice either to soil his bed or else use the litter provided and the instinct to keep the bed clean will usually prevail.

Gradually, over a matter of weeks, the area of floor covered with litter should be decreased until it is equivalent in size to an average litter tray and eventually an actual tray can be brought into the pen to hold the litter. In order to encourage Gizmo to continue to use the litter as the area diminishes, it will help to place tubs of dry cat food on the previously covered floor since cats are understandably reluctant to soil on or near their food.

In most cases a period of confinement of seven to fourteen days will be enough for a cat to establish an attachment to litter and accept the tray as a suitable latrine. Once you are confident that this has happened, Gizmo should be allowed out of the pen but given access to just the one room. The litter tray can then be moved farther and farther away from the bed and he can be encouraged to make more of an effort to locate it in order to eliminate.

The next stage is to allow access to the rest of the house, one room at a time, while ensuring that Gizmo is supervised at all times. When supervision is not possible, for whatever reason, he should be returned to the pen in order to prevent any setbacks in treatment.

As you can see, this treatment method is long and drawn-out and since Gizmo is a long-haired cat there will also be additional grooming and cleaning needed while he is confined to the pen. All this will require commitment and patience on your part, but if all goes well you can expect your precious little cat to be as strongly attached to litter at the end of treatment as any self-taught cat would be.

Hunting

See Predation, p. 178.

I

Indoor toileting

In the section on house-training we looked at the problems encountered when kittens fail to learn to use a litter tray, but I also see a great number of cases where a cat that was previously clean in the house suddenly starts to make a mess. Such cases are extremely distressing for the owners, and yet by taking the time to ask the right questions it is often possible to identify a relatively simple and often correctable cause.

Question

I am almost too embarrassed to write to you but I feel that I cannot avoid this problem any longer. The fact is that my house is beginning to smell of cat urine and sooner or later visitors are going to start to notice. Until about six months ago my cat Cuddles was the cleanest creature you could imagine, but then she had some trouble with her waterworks. She was going to the toilet all the time and I had to take her to my vet. He told me that she had an infection and gave her some tablets which seemed to do the trick. While she was poorly I didn't mind the odd accident on the carpet, but it's ages since the vet gave her the all-clear and she is still going to the toilet in the sitting-room behind the settee. Before she was ill she used a litter tray all the time and she hasn't forgotten what the tray is for because she still does her solid movements in it. I keep the tray as clean as possible and I haven't changed the type of litter I use, so why does she insist on using the sitting-room as her lavatory?

Answer

The embarrassment you express at the start of your letter is a very common reaction to problems of indoor toileting.

Owners like yourself usually tend to blame themselves when their previously clean cat starts to make mistakes, and worries about lack of hygiene add to the emotional stress of these problems. A surprising number of people actually stop having visitors because they are so concerned about the smell that is gradually pervading the whole house.

Cat urine has a particularly lingering odour, and scrupulous cleaning up is important not only for the sake of the humans in the house but also for the cat, since the smell of urine will attract it to return to places where it has urinated before and so perpetuate the problem. Thorough cleaning using a warm solution of a biological washing powder should be followed by spraying the soiled area with an alcohol such as surgical spirit, and the cat should not be allowed access again until the area has dried completely.

You mention that you keep the litter tray as clean as possible, and obviously you do so for all the right reasons. However the principle of urine odour attracting the cat to use a particular latrine applies as much to suitable receptacles like litter trays as it does to carpets, and excessive cleaning may actually discourage some cats from using the tray. Obviously a dirty litter tray will be equally uninviting and a happy balance needs to be struck, but in general cleaning the litter tray out completely every two to three days should be sufficient for the single cat. When more than one cat shares the tray, then obviously cleaning needs to be increased accordingly, and in multi-cat households, where one individual is refusing to use the tray, it is always worth considering providing extra trays since some cats simply don't like sharing.

In Cuddle's case the refusal to use the litter tray is a selective one since she is continuing to defaecate in it without any hesitation, and this is a major clue to the cause of her problem. You mention that she has been suffering from a urinary tract infection and that her problem seems to have occurred ever since her illness. One of the symptoms of cystitis, as many women will tell you, is that passing water becomes very painful. There is a distinct burning sensation as urine is passed, and this pain becomes associated with the litter tray since that is where it is felt. In response, cats like Cuddles seek alternative safer places to relieve themselves, in the belief that the tray is in some way responsible for their discomfort.

Some owners inadvertently add to this feeling of insecurity by attempting to give the antibiotic tablets to their cat while it is sitting on the tray. Their reason for doing this is that it is the only time when the cat stays still long enough for them to get a tablet down, but unfortunately the plan is more than likely to backfire since the cat will now view the tray as a place where unpleasant things happen.

Cystitis is not the only medical condition that may lead to problems of indoor toileting. Similar problems can be seen in very old cats who are suffering from arthritis, since trying to squat on the tray can become a very painful experience for them, too. In addition to the pain factor, older cats and those suffering from cystitis will have to urinate more frequently than normal and many are caught short every so often, leaving a wet patch in the middle of the carpet. Although you were happy to clear up after such mistakes while Cuddles was poorly you are understandably less inclined to be patient now that she has been pronounced fit by the vet. Unfortunately what has happened is that, after the initial accidents on the carpet, Cuddles has established an association which means that she now views your sitting-room as her latrine, and even though the original medical problem is cured the wet patches still keep appearing.

Now that we can see why you are having problems, we need to try to persuade Cuddles to revert to using the tray for all of her toileting needs. First you must re-establish a toileting association with litter in preference to carpet, and you can achieve this by using the confinement techniques discussed under House-training (p. 137). Cuddles also needs to be persuaded that she will be just as safe and secure on the litter tray as she is behind your settee, and you will find that it helps if you place the tray in an equally protected position. You may find that she is happier if the tray is covered in some way, since this will increase her feeling of security and will be more like the areas that outdoor cats select for toileting, such as under trees or behind garden sheds. There are various brands of covered litter trays available in the pet shops, but before you rush off and spend a lot of money it is sensible to see how Cuddles reacts to a covered tray—an inverted cardboard box will serve the purpose just as well. Changing the appearance and position of the tray in this way will help to reduce Cuddles's unpleasant associations with her tray and increase her confidence in using the toileting facilities you provide.

For some owners the breakdown in their cats house-training is not restricted to urination and there is a total refusal to use the tray. In these cases there is no link with a medical problem and often it is difficult for the owners to identify any significant changes that could be causing the cat's response. It may simply take another person looking at the problem objectively to spot the obvious.

One such case involved a beautiful tabby cat called Pepper who had always used her litter tray and never caused any sort of problems. She had lived with her owners, Sandy and Peter, for five years and in that time the tray had always been positioned in the kitchen although it had been moved slightly a few months before the problems began. It was this significant moving of the tray that had resulted in Pepper's out-and-out refusal to use it, and the reason became obvious as I talked the situation through. Five months before Pepper's owners had acquired a Golden Retriever puppy called Solo, and the cat and dog had become very good friends. There was absolutely no hint of Pepper feeling insecure due to Solo's arrival and in fact the two had become virtually inseparable, but ironically it was this friendship that had inadvertently caused Pepper's problem.

The kitchen in Sandy's and Peter's house was typical of most modern estate houses—relatively small with a limited amount of floor space. When Solo moved in with her rather large bean bag, some reorganising had had to be done and as a result Pepper's litter tray had been moved into the corner next to her food bowl. It was that simple move that had resulted in the present toileting problem.

No cat likes to eat next to its toilet and Pepper was no exception. Indeed, close proximity of litter trays to feeding areas is a very common cause of breakdown in house-training and one that can be easily avoided. The treatment simply involved moving the tray back to its old place, away from the food, and finding some other way to accommodate Solo's bed. This case illustrates how important it is to look at the complete picture when treating indoor toileting problems in order to make sure that some vital clue is not being overlooked.

Injections
See Fear aggression, p. 120.

Inter-male aggression
The specific situation of aggression between two male cats is most frequently encountered as a competitive scenario involving territory boundaries or the attentions of available females. The out-and-out aggression involving the full spectrum of hissing, growling, biting and scratching is usually preceded by ritualised threat displays. The cats arch their backs and, staring piercingly at one another, stand sideways on in order to increase their perceived stature. Their hair stands up like bristles and the lips are drawn back to reveal their glistening white teeth. Growling deeply, the two adversaries will approach one another, often walking almost on tiptoe, and as they become level one will initiate the attack which is usually directed at the head and neck area.

After a period of intense fighting it is not uncommon for the cats to take a break from actual physical aggression and once more engage in a staring match, presumably designed to make one or other admit defeat and retreat. If neither of them backs down they will set upon each other once more and sometimes inflict severe physical injury.

This behaviour, which is usually seen in entire males, has both inherited and learned components and is directly influenced by the environment in which the cats are living. The most reliable treatment is castration, but when the problem is seen in neutered males, as it sometimes is, drug therapy using progestins is helpful. (*See also* Castration, p. 98, and Fighting, p. 123).

Introductions
As the popularity of the cat increases, so do the number of people who wish to own more than one cat. Many will get both their cats at the same time, either as littermates or as two unrelated individuals from a rescue centre or such like, but when a decision is made to introduce a new cat to a long-standing resident, problems can result.

Question
Two years ago we got our first cat, Minty, and we have derived hours of pleasure from her ever since. In fact we enjoy her

company so much that we decided to get another cat and recently went to the local cat sanctuary to pick one out. Holly stood out from the rest somehow and we had no difficulty in agreeing that she was the one for us. Unfortunately, though, Minty does not share our opinion of Holly, and when we brought her home I thought that Minty was going to commit murder. I have never seen a cat react so violently, and when I tried to intervene Minty scratched my hands quite badly.

I have tried separating them, but as soon as I try to re-introduce them Minty just takes up where she left off. I am rapidly coming to the conclusion that Holly was better off in the sanctuary and I have even considered taking her back there, but I wondered if you could suggest any way in which we might be able to persuade Minty that she's not that bad after all.

Answer

It is always very difficult to predict whether a resident cat will accept a newcomer since each individual has a different level of social toleration. We know that cats are capable of living together, and although they hunt alone, colonies of feral cats will live happily in close social contact provided that food and other resources are in plentiful supply. The huge numbers of households that are home to two, three or even more cats is proof that domestic cats can and indeed often do live in harmony with one another; so long as food and affection are freely available one might expect newcomers to be welcome. Unfortunately, however, this is not always the case, as you have found out.

One factor that affects any cat's ability to share its territory is the amount of social interaction it experienced as a kitten. You do not give any details about Minty's origin and it would be helpful to know a little about her early life. In general cats lacking any early experience of social contact are poorly equipped to deal with others later on, whilst cats that have had plenty of contact with others while they were kittens, and have been used to sharing their resources, are able to develop the social language necessary to cope with company in adulthood.

Cats in the latter category will be better able to accept a new arrival and tolerate its presence within their core territory, but even so it may take considerable time before a true friendship is established. The resident cat needs time to weigh up the situation and will observe the way in which the owners respond to the newcomer. A calm and relaxed atmosphere will not only help the latest family member to feel at home but will also signal to the existing residents that it should be accepted.

It is interesting to note that in your case the cat you have chosen to introduce is already adult. In most cases kittens seem to be more readily accepted and this is probably due to the fact that they are less challenging both socially and sexually. Indeed, the atmosphere of competition can be minimised by neutering the newcomer as soon as possible and in the unlikely event that Holly has not already been spayed I would advise you to take her to your vet without delay.

It is all very well telling you that Minty's background may be playing a part in her refusal to accept Holly, but what has happened in the past cannot be changed. Instead you need to know how to maximise the chances of your two cats coming to look on one another as flatmates and this will involve paying careful attention to the way they are introduced.

The most important thing is for introductions to take place in a calm and controlled manner, so that Minty has no need to feel threatened and Holly is protected from any frightening experiences while she learns to cope with her new environment. The best way to achieve this is to use an indoor pen (see Den, p. 105) where Holly can be supplied with a bed and a litter tray, and can be fed and watered in complete safety for a few days. The pen should be located in the heart of the house so that she becomes accustomed to the noises and activities of the family, and from her protected position she can observe what goes on in her new home in the sure knowledge that she can come to no harm. Often it is the rapid movement of the newcomer in flight, after being frightened by the resident, that induces chasing and even aggression, and by preventing Holly from running away she is forced to stand her ground. The pen also allows Minty to investigate Holly from her protected position outside and enables her to come to terms with Holly's scent inside her home. Moving the pen from room to room will also help to establish Holly's right to occupy the whole house. Once Minty is a little calmer you should try to feed her outside Holly's pen whilst Holly is eating inside, so bringing them close together while using the reward of food as a distraction. Eventually you should reach the situation where the two cats are happy to eat side by side, albeit separated from one another by the bars of the pen.

Having got that far, the next stage is to attempt introductions without the pen and you will definitely need someone else to help you for the first few meetings. One or both cats should be restrained and it will probably help if you have some food nearby to use as a distraction if necessary. Over the next few days you should bring Holly out of her pen more and more and allow her to spend time exploring her new surroundings. As Minty becomes more tolerant, the amount of contact between the two can also increase, but you should supervise interactions for some time yet. It is important that both cats have access to safe havens where they can retreat if the tension becomes too much and realistically it will be some time before you can expect them to curl up together by the fire—if indeed they ever will.

In many cases the methods outlined above are successful in easing another cat into the household, but on occasion the resident is so determined to be an only cat that no amount of persuasion will make it change its mind.

The fact that Minty has consistently reacted aggressively towards Holly at every meeting does slightly reduce the prospect of success in your case, and you need to be realistic about the situation. You may find it helps to give Minty a short course of either a conventional or a homoeopathic sedative in order to encourage her to be more tolerant and less easily distressed, but as with all behavioural cases it is important not to rely on these aids to treatment too heavily. If after a few weeks Minty is still reluctant to accept Holly as part of the family, you have little choice but to accept her point of view and allow her to return to her solitary status.

J

Juvenile behaviour

In order to understand the behaviour of our domestic cats we need to understand the basis of their relationship with us and how they fit into our family existence. The dog is well known as a pack animal and in an ideal dog-owner relationship the human takes on the role of pack leader. We provide food and shelter but we also make the decisions and dictate the way that life goes on within the pack. In common with most domestic species the dog has been manipulated to the extent that it cannot realistically survive without us.

In contrast the cat has retained its independence and lives with us out of convenience rather than necessity. It has learned how to exploit us by offering affection and companionship but makes sure that it does so on its own terms.

This picture of an independent manipulating species conjures up the idea of a pet that milks us for all we are worth and gives nothing in return, but those of us who own cats know that nothing could be further from the truth. The bonds that we develop with our cats are just as strong as those we establish with our canine companions, but the foundations for them are quite different.

In basic terms, the relationship between owner and cat is a maternal one and we fulfil the role of mother long after our cats would have reached maturity and full independence in the wild. In fact our domestic cats retain their juvenile behaviour throughout their lives, somehow managing to be independent adult and vulnerable kitten rolled into one, and the majority switch effortlessly from one to the other.

As cats grow up and enter their independent adulthood, fellow felines become competitors for territory, resources and indeed sexual relationships, and sociable behaviour gradually diminishes. However, it is misleading to describe the cat as an asocial animal since feral cats demonstrate that they are capable of forming social relationships with one another provided that resources are plentiful. Studies of feral cat behaviour have shown that although they may hunt alone these cats can coexist in what are often extended family groups of related individuals, with queens sharing in the protection of their young and in some cases even suckling one another's kittens.

Having said that, the only time in the cat's life when it is totally dependent on other members of its own species is during the early days of kittenhood, and by fulfilling our maternal role we mimic this stage in feline development and encourage our pets to rely on us for food and security. We continue to provide warmth, love and affection, and in return our cats provide us with a means of relaxation which is undoubtedly therapeutic.

Question

I am an elderly woman living alone except for my cat Jessica who is a wonderful companion. She is always there when I need her and just stroking her makes me feel that I am needed too. As you can imagine, I have become quite dependent on my cat and I spend a lot of my time cuddling her on my knee, but recently I have noticed that she dribbles when I hold her. This is becoming something of a problem because sometimes she dribbles so much that my blouse is wet through. She also snags my clothes with her claws, but by cutting them regularly I keep the damage to a minimum. I phoned my vet to ask about the dribbling and he suggested that she might have a problem with her teeth, but when I took her to see him there didn't seem to be anything wrong with them. He then told me that this might be a behavioural problem and so I am asking you to help me.

Answer

It is good to know that your cat is such a source of pleasure and company for you. There is no doubt that in this ever more lonely society of ours the cat is an ideal companion, since it provides us with a focus for our nurturing and caring emotions and, as you say in your letter, makes us 'feel needed'.

In Jessica's eyes you are fulfilling a maternal role and by

cuddling her on your lap you are imitating the security that she experienced when lying next to her mother and suckling as a kitten. In response she purrs and rubs in an affectionate way and by kneading with her paws she is engaging in the same behaviour that she used as a kitten to stimulate milk flow from her mother's nipples. In anticipation of this milk Jessica also begins to salivate and dribbles copiously over your clothing. Some cats go one stage further and actually suck on their owner's neck, and some have even been known to suck hard enough to make red marks on the skin.

Most owners find the kneading behaviour perfectly acceptable and, like you, simply cut their cats' claws to prevent them damaging either themselves or their clothes, but when dribbling and sucking are both involved it becomes more of a problem. What we are seeing is an extension of the normal juvenile behaviour exhibited by all domestic cats but Jessica is beginning to take her dependent role too far and place too much importance on you as her protector.

It is important to rectify this over-dependence before Jessica loses her ability to cope without you since, although relatively unusual, it is possible for cats to develop symptoms of separation anxiety in their owner's absence (see Separation anxiety, p. 189). Treatment of cases like yours demands a great deal of commitment on the part of the owner, and the success will depend on your ability to distance yourself from Jessica while still maintaining a valuable relationship. It is important for you to become far less readily available to Jessica and ensure that affection is initiated by you rather than by her in future. This does not mean that you have to show her any less love but simply that you should dictate the times when your affection is available.

Try to encourage her to spend more of her time outside, developing her hunting skills and engaging in more adult forms of behaviour, and if possible restrict your cuddling sessions so that Jessica never reaches the point when she starts to dribble.

Ideally it would help if you could get someone else to feed and groom her so that the emphasis on you as her provider can be reduced. However, since you live alone this is likely to be very difficult and unless you have friends who can come in to help you, you will obviously have to continue doing these things, but do try to be far more detached in your attitude.

This is extremely difficult since you are bound to feel cruel and guilty for rejecting Jessica's displays of affection, and deep down you will probably want to respond to her. The fact that she is your main companion makes it even harder and you will need to be prepared for some difficult times.

However, Jessica will quickly learn to be more independent and you can both enjoy a new, more comfortable and certainly drier relationship! You will still be regarded as a mother figure and Jessica will no doubt lapse into juvenile behaviour, content to be stroked and cuddled, but she should stop short of expecting you to produce the milk to feed her.

L

Learned aggression

Question

My cat Bitsy is normally a very affectionate individual but he has one very distressing, not to mention painful, habit which is getting to be beyond a joke. Every night when I start to go upstairs to bed Bitsy ambushes me and grabs my ankles. I know that it conjures up a very amusing picture but believe me, it is rapidly losing its humour value. My skin is black and blue from the nightly attacks and more often than not he actually draws blood. Once he has got hold of me he bites into my skin and then kicks furiously with his back legs and scratches with his sharp claws.

This problem started about six months ago, and on the first occasion he merely tackled me at the foot of the stairs and did not do any physical damage. I interpreted his behaviour on that occasion as being a request for food and therefore went back to the kitchen and filled his bowl. He was most appreciative and I thought no more about it. However, from that time on the interceptions on the stairs became more and more frequent, and after several occasions when I stopped to feed him before continuing upstairs I realised that the attacks were getting steadily more vicious.

We have now reached the situation where anyone attempting to go upstairs late in the evening is attacked, and recently one of my visitors was bitten very badly on the ankle when trying to go to bed.

I have noticed that this behaviour only happens in the evenings and anyone can go upstairs perfectly safely during the day. The attacks are always short-lived and Bitsy calms down very quickly when you feed him. In fact it is even possible to prevent the attacks by simply feeding Bitsy before attempting to go upstairs, but I don't always remember and in any case I feel that his behaviour is beginning to get out of control. I am worried that he will do someone a serious injury one day if we don't try to cure him of this terrible habit.

Answer

Bisty has got you exactly where he wants you. He has discovered that by attacking you on your way to bed he can make you give him a late night snack and he has unfortunately got the message that aggression pays. On that very first occasion he may well have been genuinely hungry and tackled you on the stairs to get your attention, or he may simply have been playing. Whatever his motivation, the outcome of his behaviour was a meal and he rapidly realised that he could manipulate you into doing something for him.

The problem is that when aggression is seen to be effective in this way the animal incorporates it into its behavioural repertoire as learned aggression and resorts to it more and more readily to get what it wants. Having established that you will respond to his attacks by feeding him, Bitsy has now generalised his behaviour and attacks anyone who dares to go upstairs without feeding him first.

The good news is that most learned behaviours can be unlearned so long as the provision of reward can be controlled by the owner. What you need to do is ensure that Bitsy's attacks fail to produce food and instead result in an unpleasant reaction. Do not be tempted to use physical punishment to teach Bitsy a lesson since this will only serve to heighten his reactions and increase his tendency to be aggressive. It will bring more risk of injury and is therefore counterproductive.

Instead you should use some form of aversion technique so that Bitsy perceives the unpleasant reaction to be a direct consequence of his own actions rather than associating it with you. A water pistol rapidly and accurately aimed will do just that, and provided you can ensure that each and every 'attack' is met with an unexpected shower and never with a full feed bowl, the prognosis for reform is good.

Litter preference

We have already discussed the importance of paying attention to the position of the litter tray if we are to avoid problems of indoor toileting. It is equally vital to ensure that the contents of the tray are acceptable to our feline companions if we are to expect them to use the facilities we provide. For the majority of cats the litter tray is only needed until they are old enough to go outdoors and select their own substrate and the majority transfer readily to using garden soil. The finer the soil the better they like it and this explains why so many cats wait until their owners, or worse still their neighbours, have carefully hoed their gardens and broken up the fine topsoil before leaving their mark. Many a gardener, feline enthusiast or not, has cursed the cat that digs up his precious plants and then defeacates in the freshly tended ground.

Question
Recently we decided that it would be nice for our two-year-old son to have a sand-pit to play in and my husband made one in the back garden. Much to my disgust I find that my cat has taken to using it as a big litter tray and, judging by the amount of mess, the rest of the local cat population must be doing the same. Why should they do this and how can I stop them in future?

Answer
Although you may find your cat's behaviour rather distasteful, he is in fact only doing what comes naturally. All cats prefer to use a fine grain substrate for their toileting needs and when someone is kind enough to provide them with sand they will undoubtedly use it. The cat's ancestors came from an area of semi-desert conditions and sand is the ideal litter as far as the cat is concerned. Anyone who has built an extension on their home will have found that the sand on their drive rapidly becomes adopted as the feline public toilet, in much the same way as your son's sand-pit has done. Even if the sand-pit were to have a label on it, unfortunately the cats cannot read and they would still be totally unaware that what they are doing is causing such a problem.

However, it is most definitely a problem and one that must be sorted out since cat faeces in a child's sand-pit are a very real health hazard. Your cat may be completely up to date with his worming, and indeed he should be, but there is no guarantee that the same is true for the other cats. Obviously the sand that is in the sand-pit at the moment has been soiled and should be removed and replaced before your son is allowed to play there again. Having done that, the treatment for this problem is to cover up your son's sand-pit whenever it is not in use, so that the cats cannot get into it. You may have noticed that the modern plastic sand-pits that you can buy in the shops have lids on them, and that is precisely because of the potential problem with cats.

For the cat that lives permanently indoors and relies on a litter tray for all its toileting needs, the choice of litter is a very important one. All litters must of course be easily rakeable in order to enable the cat to dig a latrine hole and then cover it over after use. Just as the outdoor cat will seek out the fine topsoil in the garden, studies into litter preference have shown that cats will actively select a finer grain material if given a choice. In general, most cats simply accept what is offered to them and most owners give little thought to the type of litter they buy, but when problems arise it is always worth looking at the contents of the tray.

In this country the two main types of cat litter have traditionally been Fuller's earth and wood-chip pellets. The first of these, which is a clay-based product, is probably the most popular and these pale-grey granules are what the majority of people think of when you talk about cat litter. The alternative wood-chip pellets are increasing in popularity because they are substantially lighter to carry and also because they claim to be more successful in quelling odours and keeping our houses smelling fresh and clean. Many of the products even go as far as incorporating pine scent in order to counteract unpleasant cat smells and make the product more attractive to the purchaser.

More recently a third variety of litter has made its way onto the market in this country and has even been the subject of television advertising. These products, which have proved very popular on the American market, combine the advantages of being fine grain and lightweight and in addition claim to be far longer-lasting than the alternatives. The idea is that the litter forms into clumps when it becomes wet, and provided that these damp clumps are removed regularly together with any solid excreta, the rest of the litter can be left and complete cleaning of the tray can be carried out less frequently.

One thing we do need to remember is that indoor cats develop strong associations with their litter material, whatever it may be, and if it should become necessary to alter the litter type you must be careful how you manage the changeover process. Sudden transfers from one variety to another may make sense to us, either for convenience or for financial reasons, but failure to consider the feline point of view can result in cats refusing to use trays and choosing less acceptable places to relieve themselves. Any changeover should be made gradually, so allowing the cat slowly to come to terms with the new texture under its feet—and the new odour if the change is to a scented variety.

We may think that pine is a pleasant smell but some cats may be less than impressed by such added features. The smell of their own urine will help to encourage them to go back to the same latrine and if the litter produces a strong alien scent they may find this extremely confusing. In some cases a strange scent may completely deter the cat from using the tray, and it may even help to transfer some of the previous day's damp soiled litter onto the new material in order to reassure it that this is the place to go.

Question

I have three cats who all use a litter tray and as you can imagine I go through a fair amount of litter. Having spent years nearly breaking my back carrying huge bags of Fuller's earth back from the shops, I decided to try those lightweight wooden pellets instead. Two of my cats accepted them without any problems but, Kiri, my sixteen-year-old Siamese, started to refuse to use her tray as soon as I made the change and has taken to urinating under the table on the dining-room carpet instead. On a couple of occasions I did see her going to the tray, and once or twice she used it for her solid motions. However, more often she deposited these just next to the tray. My too younger cats go outside during the day and use the tray at night but Kiri rarely if ever goes out these days, and if she refuses to use the tray it is more of a disaster.

In the end I decided that I was expecting too much of my old girl to make her change litter so late in life and I changed back to the heavier variety. However, despite this she has continued to use the dining-room carpet and I am at a loss what to do now. None of my cats have ever been dirty before and I just wish that I had never heard of lightweight litter.

Answer
It is interesting to note that two of your cats have accepted the litter change, but there is little doubt that Kiri's toileting problems are a direct consequence of your efforts to save your poor back. The fact that she is elderly and is now a totally indoor cat may explain her refusal to use the wooden-type litter. Her pads are likely to be much softer than her younger friends who go outside to explore, and the texture of the pellets could well be uncomfortable for her to stand on.

It is worth noting that she returned to the tray when she needed to defeacate, but it seems that she could not bring herself to actually stand in the tray to relieve herself. The smell of urine and faeces from the other cats was probably attracting her to the right place, but once she got there she found the facilities weren't acceptable.

In view of Kiri's age you have probably made a wise decision to abandon the pelleted litter, but unfortunately in the meantime she has transferred her substrate preference to carpet and adopted your dining-room as her latrine. It is likely that the smell of a feline latrine is as important as its position and contents, and therefore thorough cleaning is vital in order to break unwanted associations. Many household cleaners are totally unsuitable for this job since they are ammonia based and will only serve to confuse the cat further (see Spraying, p. 191). Instead, any areas used by Kiri should be cleaned using a warm solution of biological washing powder followed by surgical spirit, and left to dry thoroughly before she is allowed access to the dining-room again (see Indoor toileting, p. 141). In some cases, when the smell is particularly well ingrained, discarding the affected piece of carpet may be the only way to remove all odour effectively and ensure treatment success.

The second phase of treatment will involve re-establishing Kiri's latrine association with cat litter, and you will best achieve this by confining her either in a small uncarpeted room or an indoor pen. If you line the litter tray with a small piece of carpet she will be encouraged to use it, and once she is happy to do so you can start to reintroduce her to litter. At this stage you may wish to consider your choice of litter and if you want to avoid carrying those huge, heavy bags again you could always try the lightweight fine grain variety. Begin by sprinkling a very small amount of litter over the carpet in the tray and then gradually increase the thickness of the litter until eventually the carpet is completely obscured. After a few days, when Kiri is using the tray reliably, you should remove the carpet altogether and then gradually allow supervised access to other rooms in the house, always ensuring that a tray is readily accessible. Unsupervised access to carpeted rooms and the dining-room in particular should be delayed for a couple of weeks in order to minimise the risks of a breakdown.

Finally, if you ever decide to change your brand of cat litter again, for whatever reason, do make sure that you make the changeover very gradually and do not be tempted to clean the tray too often since the scent of urine will help to attract Kiri and persuade her not to go looking for alternative facilities.

M

Marking

Communication in the feline world is a complex business and marking is a vital part of it. There are various ways in which a cat may put his mark on his surroundings and all are perfectly acceptable in their place. However, what a cat would consider to be a legitimate marker can sometimes be quite offensive to the uninitiated human, and when cats start to mark inside the home, relationships can become quite strained. Whereas the sight of a cat gently rubbing itself along your hi-fi unit may be endearing, it is quite another matter when the marker being used is urine and your cat starts spraying on your toaster! Similarly, scratching the fence posts in the garden is no cause for concern, but sharp claws pulling at an expensive sofa are rarely accepted willingly. There are four basic methods of marking used by the cat—rubbing, scratching, spraying and middening—and each of these is dealt with in the relevant alphabetical section.

Maternal aggression

The aggression displayed by queens who are nursing their offspring is probably the most understandable and therefore accepted form of what is otherwise an undesirable behaviour. It is a behaviour pattern that is linked to hormone activity and is triggered by the presence of young kittens. According to behaviourists, this is one of the few occasions when a female cat will behave as a true aggressor, and indeed maternal aggression can be one of the fiercest forms of attack ever experienced in the feline world.

The most distressing aspect of this behaviour, as far as owners of domestic cats are concerned, is that the female may attack both strangers and previously well-trusted family members with equal venom. Often she will give no threatening display as a warning of her intentions, but rather launch herself at the approaching challenge in an almost eager manner; even the most placid of cats may lash out uncontrollably if anyone attempts to remove her kittens from their nest.

Such behaviour is not only exhibited towards people in the house who come too close to the precious offspring, but will also be invoked by other cats or indeed other household pets who dare to show too much interest in the new arrivals.

One explanation given for the ferocity of maternal aggression is that the queen is prevented from exercising her usual flight response to any perceived danger because she must remain with her offspring and protect them. As a result she may become increasingly agitated and excitable and her aggression threshold is correspondingly lowered. In this situation any threatening gesture, however unintentional, will invoke an explosion of anger and an often quite frightening attack.

Question
My young daughter has always been fascinated by animals and shown more than the usual childhood interest in all our pets. Even though she is only ten years old, she has already decided that she wants to work with animals and if possible she would love to be a vet. She is very keen to get as much experience of handling and looking after animals as possible, and when we got another cat last year she immediately pestered me to let her have kittens. Her argument was that if her own cat had kittens she would be able to see the whole process of kitten rearing at first hand, and it would be valuable experience.

Unfortunately, however, this has not been the case because our normally docile and loving cat has turned into a vicious and unsociable individual and is adamant that no one will get anywhere near her kittens. My daughter has always regarded Snowball as something of a special friend and, as you can imagine, she is most upset by her rejection. She is a very gentle child and has not given the cat any reason to think that she would harm the kittens, but it seems impossible to convince Snowball that we only want to be friends. If we were to let her have another litter in the future, is there anything we could do to prepare her for the birth and make her happy to let us share in the joy of her kittens?

Answer
Often cat owners do find that their pets allow them to be very closely involved with their kittens, and we regard it as a great privilege to watch the kittens develop. However, we can never expect to claim such involvement as a right and we must always wait to be, as it were, invited. Obviously it is very distressing for a child to feel rejected by a pet that she dearly loves but you need to explain to your daughter that Snowball's aggression is not a personal reaction against her, but rather a natural feline behaviour dictated by hormones and largely outside Snowball's control.

It is thought that maternal aggression may be related to the falling levels of progesterone following the birth, and very extreme cases may be helped by the administration of artificial progesterone-related drugs. However, in general the best treatment is simply to provide the queen with a quiet nesting area where she will be undisturbed, and refrain from attempting to handle the kittens in her presence until she has become more relaxed and accepting of your company.

On the subject of any further litters which Snowball may produce, there does not seem to be any guarantee that the aggressive behaviour will be repeated and she may indeed allow you to be far more involved next time. However, even when queens do not display aggressive behaviour it is always important to be cautious in approaching and handling young kittens. I am sorry that your daughter did not obtain the insight she would have liked into feline reproduction, but she has most certainly been given a valuable lesson in feline behaviour.

Mating
See chapter 5, under 'M'.

Middening
This is the term used to describe the depositing of faeces in chosen sites as a deliberate marker. Many species of wild cat are known to leave their faeces uncovered in conspicuous locations and it is believed that they are used as a source of scent-coded information, but the specific role of middening in

feline communication is still unclear. Other species such as the otter are known to deposit their excreta in discrete piles on the tops of rocks, and these 'spraints' as they are called are used to define territory boundaries.

The incidence of middening within the feral cat population appears to be greater in areas of high population density, which may result from the increased risk of threatening confrontations and a need to establish rights over territory. Certainly, dominant cats seem to use faeces in this way far more than their more subordinate counterparts, and the 'messages' are usually left on elevated points inside the territory. It is likely that domestic cats living in highly populated urban areas use middening as part of their communication network and deposit faeces as a signal to their neighbours when territory rights are in dispute. The aim of such marking is much the same as it is with spraying, but perhaps faeces act as a more blatant reminder of a feline presence.

Just as spraying in the outdoor environment is a perfectly acceptable pastime for our cats, so middening is unlikely to cause any problems unless these little parcels begin to be deposited at strategic points inside the house.

Question

Our cat Oliver has always been very clean, but ever since we went away on holiday and left him in the house on his own he has started to do his solid toilet just by the back door. He was checked on regularly by our neighbours while we were away and we have a cat flap in the back door so that he had no excuse for not going outside. There was no evidence of any urine in the house, but my neighbours had to clear up the other mess virtually every day.

Since I've been back I have even tried putting a litter tray in the house, thinking that perhaps he was just unwilling to go out in the cold, but he simply ignores the tray and continues to go by the back door. Is this some sort of protest at our going away and if so, why doesn't he stop now that we are home again?

Answer

The position of these little presents from Oliver and the fact that he has continued to be clean in other ways suggest that this is more likely to be a case of middening than a breakdown in house-training. While you have been away Oliver has not been able to rely on you for security and he may even have been subjected to challenges from local cats. The fact that the faeces are deposited by the back door which also houses a cat flap is highly significant. This is an area of challenge where scents are frequently changing, and invading cats may have entered through the flap, bringing with them their own scent and special sort of conflict.

Faced with such a challenge to their security, some cats might have begun to spray their indoor territory, but Oliver has chosen to make his mark in a more obvious way. Often middening is chosen in preference to spraying when the challenge is greater or the cat concerned is less emotionally stable, and both of these factors may make treatment more difficult to effect. Many people find the sight of cat faeces particularly offensive, and although feline urine may be quite pungent it is often better tolerated because it is not visually disturbing!

Treatment is aimed at increasing Oliver's general level of competence and must be applied with patience and understanding. The use of an indoor pen in order to confine him while he learns to cope with challenge is recommended, and boarding up the cat flap temporarily if not permanently may also help to reduce his feelings of insecurity. Restricting Oliver's access to one or two rooms in the house will enable him to increase his confidence and he can gradually be given more freedom.

Where middening has occurred in response to an identifiable challenge such as moving house or introducing a new cat, then techniques of controlled exposure should be employed to reduce the cat's sensitivity to the stimulus (see Introductions, p. 145). In Oliver's case, I would strongly recommend that when you go on holiday the next time you put him into a good local cattery rather than leave him at home where your absence could once again present an unbearable challenge to his confidence (see Boarding, p. 92).

N

Nervousness

In any species, part of the learning process of growing up is learning how to distinguish between threatening and harmless stimuli. All kittens display the same inherent reaction of flattened ears, arched back and fluffed-up coat when startled by a sudden movement or alarming sound. This frightened feline posture is utilised in many a cartoon and we are all familiar with it from watching our share of *Tom and Jerry*. However, as the kitten matures and finds itself exposed to an ever-increasing variety of environmental stimuli, it will gradually learn to adjust to challenge and accept novelty. It becomes more accepting of the world around it and is able to deal with new experiences without necessarily resorting to a fearful response.

As time goes by, the habituation process, as it is called, results in the development of a competent cat well equipped to cope with whatever the world has to offer, but it never loses that inherent startle reaction and will demonstrate it in particularly alarming situations. Once this competence has been acquired it will be continually reinforced by the on-going exposure to challenge throughout life, and most cats will develop into easy-going and adaptable pets.

However, for some, life is a continual source of anxiety and they never quite learn how to face up to their environment. These are the cats that spend their time hidden in the darkness beneath the sideboard, looking out into the room with widely dilated eyes and desperately hoping that no one will notice them. Once they feel secure in their dark hole they will lie low until the perceived challenge has passed. In this way they aim to live out their days, avoiding challenge and carrying the 'head in the sand' approach to its extreme. The threshold of fear is extremely low in these cats, and rather than sitting motionless in the corner some will take flight at the smallest alteration in their surroundings, often flying out through the cat flap, not to be seen again for some hours. They do not wait to weigh up the situation but work on the principle that all change should be avoided and therefore they never learn to face up to new experiences or cope with even the mildest of challenges.

Whichever reaction the cat chooses to adopt, one thing is certain—life for these cats is a distressing experience. For some the situation becomes so severe that they begin to overreact to all the harmless everyday events within the family and lose hope of ever enjoying life to the full. The owners are usually equally distressed and find it hard to sit by and watch as their much loved pet shrinks farther and farther into its shell. Often they try to approach the cat with reassuring gestures and even attempt to pick it up and cuddle it to prove that the world is a friendly place. Unfortunately such comfort is difficult for the cat to accept in its state of intense anxiety, and the fact that the owner is drawing attention to him when all he wants to do is hide can be interpreted as yet another challenge. In some cases the response is a display of fear-related aggression and a badly scratched owner, all of which serves to further reduce the threshold of the cat's nervous reactions and in reality make the situation far worse.

In order to understand why some cats react so badly to simple challenge it helps to take a look at the early days of kittenhood and the way in which they were introduced to the big wide world. Behaviourists have shown by extensive research that the frequency of handling experienced by young kittens between two and seven weeks of age has a marked influence on their ability to appreciate and enjoy human contact in adulthood. Similarly, the range of environmental stimuli that the kitten encounters during the first few weeks of its life will affect its ability to react positively to the many opportunities of life and learn that novelty can be rewarding rather than threatening. It has also been shown that the temperament of the queen can affect the nervous disposition of her offspring: in general, inexperienced and incompetent mothers produce a further generation of nervous and withdrawn individuals.

All of this is true and must be taken into account, but life

Life for a nervous cat is a distressing experience.

is never that simple and there is also a wide variation in tolerance thresholds between individuals, regardless of their parentage or early experiences. Just as no two children within a family react in exactly the same way to their environment, so kittens from the same litter will be capable of exhibiting quite different responses in the same situation. The bolder and more extrovert characters will be eager to make the most of life and will investigate all new stimuli with obvious enthusiasm, while the smaller and less forthright will tend to hold back and wait until everyone else has had a look before venturing forward to greet a novel challenge. In some cases they may even choose to avoid the situation altogether and, despite the fact that these kittens have been offered exactly the same opportunities in early life, they will grow up far less competent than their adventurous littermates.

For some cats, their nervousness is generalised and stems from an inability to cope with life, but for others their fear is far more specific and may be described as a phobia. Such cases will often derive from a single traumatic experience and are likely to develop later in life. True phobias are very rare in the cat, but where they do occur treatment is similar to that of the nervous cat and involves controlled exposure to the specific challenge (see Agoraphobia, p. 76, and Xenophobia, p. 212).

Question

I hope that you will be able to offer me some advice regarding my cat Comet who is becoming increasingly unable to cope with life. He is four years old and, if I am honest, he has always shown a tendency to be rather nervous, but instead of improving with age it is steadily getting worse. I live alone and quite regularly go to stay with my sister and her family who live some distance from my home. As a result of Comet's rather withdrawn character I have always been reluctant to put him into a cattery, and since my sister is also very fond of cats I take Comet with me when I go away. However, he does not seem to appreciate the hospitality and from the moment he arrives he slinks off under the sideboard and refuses to come out. If I put his feed bowl under the sideboard too, he will gladly eat, but if the bowl is put in the open he waits until we have all gone to bed before he ventures out. It is obvious that he is not enjoying his holiday but I worry that he would be equally miserable if I left him behind. What should I do?

Answer
Poor old Comet reminds me of the child who spends the whole of the birthday party locked in the toilet but goes home and tells his family what a great time he had! Certainly Comet would be classed as a nervous cat and his behaviour must be distressing for you to watch. Obviously your reasons for taking him with you when you visit your sister are sound, but unfortunately they have rather backfired on you.

It is interesting that you describe Comet as a 'withdrawn character' and say that he has been so ever since kittenhood. Here we have a cat who never fully learned to cope with life, and in common with all such cats he has gradually deteriorated over the years because of his avoidance of any potential challenge. By trying to protect him you have simply enabled him to go through life without having to face up to his fears, and without treatment the situation will steadily worsen. Despite the fact that your sister loves cats, Comet is still faced with what he perceives to be a hostile environment and, true to his character, he reacts by withdrawing and hiding in a conveniently dark and quiet corner. He is used to living with just one person in what must be relative peace, and your sister's home with a family around is bound to offer far more challenge than Comet can bear. Often cats like Comet will become very secretive in their behaviour and indeed, the fact that he is willing to wait for his food until you have all gone to bed and the coast is clear is further evidence of just how severe his condition has become.

Treatment must be aimed at helping Comet to face up to life and so learn that it can offer many rewarding experiences if he would only come out of the 'closet'. Obviously it is far easier for a cat to adapt while it is young, and when nervous cats have already reached adulthood before treatment is begun they will rarely transform into totally competent felines. However, it is possible to make vast improvements and enable cats like Comet at least to begin to enjoy life once more.

As with any nervous creature, the approach to treatment must be calm and gentle and a great deal of patience is required. Never be tempted to steam ahead with the treatment programme but rather gauge your speed of progress by what Comet is able to accept. Controlled exposure techniques using an indoor pen will enable Comet to experience normal everyday life and prevent him from trying to escape. From inside the pen he can observe the comings and goings in the sure knowledge that he can come to no harm, and so protected he can afford to face up to challenge.

In your case I would suggest that you work hard to increase Comet's competence in his own familiar surroundings before you take the pen to your sister's house and start to habituate him to his holiday home. If you find that at first he reacts frantically to being confined in the pen, his fear response can be temporarily quelled by the use of sedatives or homoeopathic remedies after careful discussion with your veterinary surgeon, but do not be tempted to rely on such treatment since Comet must learn to cope without becoming dependent on medical assistance.

Neutering

See Castration, p. 98, and Spaying, p. 190.

O

Old age

One of the advantages of the excellent service that the veterinary profession provides for pets in this country, and the ever improving medical and surgical techniques that are available, is that our pet cats live longer. Over recent years there has been a noticeable increase in the number of cats in their late teens, and reports of cats reaching their twenties are no longer regarded as particularly remarkable.

However, just as old people make special demands on those responsible for their care, so old cats necessitate a degree of extra tender loving care and special consideration. Most of us are so thankful that our dear companions are still with us that we make generous allowances for them if their behaviour is not always as we would wish, but there may be occasions when an old cat presents behaviour problems that stem directly from its advancing years.

In general, older cats are less able to cope with the hustle and bustle of a busy family, and when young children are around they must be taught to be gentle with their ageing pet and give him the respect that his years entitle him to! All cats, but those that are getting on a bit in particular, should have access to a safe and secure haven where they can escape from the activity and sleep their days away in peace and tranquillity.

It is only natural that as their body begins to slow down so they will need to sleep for longer periods and will spend increasing amounts of time indoors. Even the most proficient hunter will start to reduce the number and length of his expeditions as the teenage years approach, and although few will ever relinquish their independence completely, many do appear increasingly willing to take advantage of the protection that we offer.

Many old cats begin to suffer from arthritic changes in their joints and become less and less agile as the years go by. One of the consequences of this is that they are unable to walk great distances and will be less inclined to jump up onto high surfaces. It is therefore important to provide a sleeping place for them at a low level and also to ensure that their food and water are always readily accessible.

Similarly, the older cat which is now spending more time indoors may need the provision of a litter tray and this must be made easily available. The restrictions of arthritis and the consequences of a weakening bladder sphincter may mean that your old age pensioner often has to get to his tray in something of a hurry and it will help to provide a number of trays in strategic points throughout the house, particularly close to his favourite resting places. This will help to avoid the problem of accidents on the carpet when he wakes up and needs to relieve himself but his poor old legs won't get working quite fast enough.

Old age often brings with it decreased levels of patience and lower tolerance thresholds, so your geriatric pet may appear slightly more quick-tempered than he was in his youth. This is a side-effect of the passing years and is recognised in the human species as much as in cats! Basically we need to give a little and try not to pressurise our cats with excessive and unwelcome stimulation, while still making them feel secure within the family unit.

An increasing demand for physical contact on their terms is another hallmark of the ageing feline, and gentle grooming is a good way of helping to keep his coat in good order while providing the love and attention that he requests. Some cats will become noticeably more vocal and seek to gain your attention by yowling pitifully, especially during the night when they are likely to feel most vulnerable. Many owners worry that such cries are an indication of pain, but provided that you have had your cat checked by a vet and he has been given a clean bill of health, such concern is unnecessary.

Finally, do remember that old age will bring with it a greater tendency to physical illness, if only because the body is beginning to wear out. Regular visits to the vet will help to identify diseases such as kidney failure at an early stage and allow treatment to be instituted as quickly as possible. It is also important to appreciate that the dietary requirements of a geriatric are not the same as for an active young adult and your veterinary surgeon will be able to advise you on how best to feed your ageing cat.

P

Pain-associated aggression

We all recognise that aggression is a common consequence of a painful experience and many of us will have reacted angrily at some time or other to our own pain. When that pain is inflicted by another party our aggression is likely to be targeted at them, and not surprisingly our cats may react in much the same way. Any of us who has quite accidentally shut our poor cat's tail in a door will have felt the force of his indignant rage and may well have the scars to prove it. Such episodes of pain-associated aggression are usually one-off events and do not have any permanent effect on the relationship between the cat and its owner. We may feel immensely guilty for causing the pain but it is unlikely that the cat will bear a permanent grudge, and his understandable outburst can be quickly forgotten. No one is surprised when a cat that has had its tail pulled by an over-enthusiastic child turns to lash out in return, and so it is wise to ensure that cats and young children are never left together unattended.

These examples of understandable aggression result from specific incidents, but life is full of more minor painful experiences, and it would be most inappropriate for every one of these to be greeted with a display of uncontrollable aggression. While still in the nest, young kittens learn to control their responses to pain through mock fights with their littermates and come to realise that aggression is not an appropriate response in all situations since it may invoke a painful counter-attack from the opposition. When kittens are denied these early experiences they are far more inclined to resort to aggression and for this reason, kittens reared in isolation rarely make suitable pets.

Pain-associated aggression which is encountered in the

elderly cat is often the result of the chronic nagging pain of arthritis which causes on-going irritation and lowers the cat's tolerance to being handled. Such a cat merely wishes to be left in peace, and indeed this is the best treatment in such cases (see Old age, p. 171). Other examples would include the convalescing cat who is suffering pain as a result of an injury and reacts aggressively to those who are attempting to nurse her back to full strength. Once again the reaction is understandable and we expect to have to show a certain level of patience and tolerance with these cats.

The treatment for the first examples of pain-associated aggression given above is basically to avoid inflicting the pain in the first place, and for the later examples to avoid unnecessary and over-enthusiastic handling of cats that are obviously in pain. This sounds simple, but for some owners pain-associated aggression is encountered as a result of a necessary interaction with their pet and avoiding the situation is not a practical solution. These are the cases of uncontrollable aggression in response to being groomed and they are encountered most frequently in the long-haired breeds.

Question

I was once the proud owner of a very beautiful little Persian kitten and I had looked forward to the day when she would be an even more beautiful adult Persian cat. Unhappily that situation never arose since she makes it impossible for me to groom her, and instead of having a magnificent mass of long silky hair she now has a disgusting mass of matted fur. I have taken her to the vet on a few occasions and had her put under anaesthetic so that they can cut out the mats, but I am concerned about having to do this too often because of the risks of multiple anaesthetics, not to mention the cost! I try to be very gentle when I groom her, but I only have to catch the comb on a tangle and she flies at me with her teeth and claws and causes quite nasty injuries to my hands and arms. I have tried wearing protective clothing but it's difficult to hold a comb, never mind use it, when you're wearing thick gloves and sleeve protectors! I am reaching the stage where I am considering having her put down if there is no way to overcome this problem.

Answer

Long-haired cats may look very beautiful on television and in books, but owning one is most certainly a time-consuming affair. If they are to be kept in their 'beautiful' condition they must be groomed daily, and their owners need to be dedicated to the cause. Even when this is the case, as no doubt it is with you, sometimes the cat is far from co-operative and we see countless numbers of these cats coming into the surgery to be regularly dematted under anaesthetic.

The problem in these cases often begins as one of pain-associated aggression, but as time goes on a learned component also becomes incorporated into the behaviour. These cats will often begin to struggle and bite furiously as soon as their owners try to restrain them and do not wait until the comb meets a tangle before they complain.

Obviously you cannot simply give up and let them get away with it unless you are prepared to have them periodically clipped and leave them to get filthy and matted in between. Prevention of the problem would most certainly be preferable to cure in these cases, and breeders of long-haired cats should be investing time in conditioning their kittens to accept being groomed from the first time that they are handled.

However, such advice is a little late for the likes of your cat and now she needs to be encouraged to accept your efforts while ensuring your safety at the same time. Cat muzzles are not commonly used and many people are unaware of their existence, but they can be most useful in treating cats who will not tolerate being groomed. Unlike the canine muzzle, this piece of equipment does not hold the mouth closed but rather acts as a hood over the cat's face and it works on the same principle as blinkers on a horse. With one of these muzzles fitted the cat will quieten and flatten itself onto the table, and with your subject in this subdued state you can slowly and gently begin to introduce her to daily grooming.

Phobias

See Nervousness, p. 165.

Pica

The definition of pica is a depraved appetite leading to the ingestion of non-nutritional substances and the most bizarre example of it in the feline world is the behaviour of wool eating which is dealt with as a separate heading (see p. 207). However wool eating is not the only example and cats have been reported to eat a variety of other peculiar materials, including rubber and even electric cables. The motivation for these behaviours is as yet unclear, but treatment involves making these items unpleasant by the application of suitable taste deterrents such as eucalyptus oil and then redirecting the cat's attentions by offering increased stimulation either through play or increased access to the outdoors. Adjusting the diet may help in some cases, even though there appears to be no nutritional component to the behaviour.

Question

I have never been one for keeping houseplants and in fact I have never been able to keep them alive for long, but I have recently got married and my wife is intent on turning our home into a greenhouse! Ordinarily this would not be a problem, but I have brought to this marriage one very loved cat—at least loved by me. The problem is that Freebie has taken rather a liking to these houseplants and has decided that they make a very nice little extra in addition to his cat food. Please can you suggest how I can stop him from devouring my wife's precious plants, before I qualify for the *Guinness Book of Records* for the world's shortest marriage?

Answer

Plant eating and grass eating are probably far more common in cats than we realise, but fortunately most of the plants that our cats choose to devour are out in the garden and we are none the wiser—unless of course they start to eat next-door's precious rare garden plant and get caught in the act! It may be that this vegetable matter is consumed as a source of roughage or that it provides certain minerals and vitamins which are not available in other foods. Another possible explanation is that plant material has some emetic properties and assists the cat in bringing up hairballs from its digestive tract.

Whatever the reason, plant eating is in fact normal behaviour even for the carnivorous feline, but such information is unlikely to placate your irate wife. The use of aversion techniques such as a water pistol may be useful to make consuming houseplants an unpleasant experience, but in order to prevent Freebie from devouring the entire collection it would be sensible to provide him with his own plants. Many pet shops sell tubs of seedling grass sprouts specifically for this purpose and most cats find them far more attractive than the average houseplant and are impressed with their owner's thoughtfulness in providing such a delicacy!

Predation

Hunting is obviously accepted as an integral part of the feline behavioural repertoire, but having said that, there are aspects of this natural predatory function which some owners find less than acceptable. For those cats that lead a double life and spend a significant portion of their time away from home, hunting is a popular pastime. They do not need to hunt and kill in order to feed themselves since their owners provide all their nutritional needs out of a can, but hunting for the cat is as much for pleasure as for necessity and so it continues to be practised by even the best-fed pets.

The sight of a cat stalking its prey is an awe-inspiring one and its precision and speed are certainly to be admired, but for the owners of the most proficient feline hunters the toll on the local wildlife population can be cause for concern. Many owners have asked me how to curb this bloodthirsty sport and have told how their intelligent cats have overcome problems such as bells on the collar by learning to hold their necks in such a way that the bell makes little or no noise (yes, we are dealing with a very adaptable species).

The short answer to such enquiries is that there is little anyone can do to dampen what is, after all, a well entrenched natural urge to hunt. Once a cat has developed into an efficient killer the behaviour cannot be eliminated, but it may be possible to lessen the effects of his murderous tendencies. Bells attached to the collar may not work for all cats but in many cases they do help to give the victims prior warning of the approaching predator and afford them the opportunity to escape.

Slight adjustments to the cat's routine may give the wildlife more of a chance if, for example, he is kept in at dawn and dusk when small mammals are most active and vulnerable. Altering the diet and offering food which demands more effort to be consumed, such as meat still on the bone, may help by increasing the time spent devouring meals and leaving less time or energy for the supplementary hunting.

Keeping the cat as a permanently indoor pet is not an option if it has been used to the freedom and independence of an outdoor life, since such enforced confinement may lead to unbearable frustration and the development of far more undesirable behavioural problems. On the other hand, for an owner who wants to prevent hunting before it has begun, an indoor pet may appeal and it would be sensible to select a kitten from a non-hunting queen in order to decrease the level of learning from its mother's example.

However, even by using these measures it will only be possible to modify the hunting behaviour rather than eliminate it, and if these cats are later given the opportunity to get out and about many will rapidly develop into serial killers.

Question

I reluctantly accept the fact that there is little that I can do to deter my beloved cat from killing birds and small rodents, but must he bring them home and leave them under our dining-room table? Sometimes all I find is the head and part of a tail, but on occasions the poor victim is complete and Rags has not made any attempt to eat it. It almost turns my stomach to find these presents first thing in the morning, but the final straw came last week when I came downstairs to find a mouse running about in my dining-room and Rags playing a macabre game of hide-and-seek. Why should my otherwise adorable cat be so cold and insensitive?

Answer

The simple answer to your final question is, because he's a cat! Although you may find this behaviour 'cold and insensitive', Rags is merely being true to his genetic make-up and behaving as the murdering feline he really is. Often we find it difficult to reconcile this image with the peaceful cat asleep by the fire, and although we can accept that they do hunt we would rather that they did not bring home the proof of their distasteful activities. However, by bringing their prey back to the house our cats are telling us that they consider our homes a safe and secure den worthy of being used as a feeding lair, and if we put aside our horror at the death of innocent victims, we can afford to be somewhat flattered by their actions.

If you feel unable to take this view and desperately want to stop Rags from bringing home his spoils, then you need to alter his perception of the dining-room as a place to consume prey. This can sometimes be achieved by using aversion techniques. Every time Rags comes home with some little parcel in his mouth he should be greeted by a well-aimed water pistol and a warning hiss, whereas when he comes home alone he should be warmly welcomed and fed immediately. This will serve to maintain the image of home as a secure base but indicate that it is not a safe place to bring your kill.

It would appear from your letter that part of your disgust with Rags stems from the fact that he is obviously not killing for food but rather for pleasure. Admittedly the sight of a bodiless head beneath the table is not a pleasant one but somehow it is easier to forgive Rags for his crime if he has eaten the victim! The fact is that most domestic cats do not need to hunt. They are well supplied with cat food and treats from their doting owners and when they return home after one of their expeditions they will often abandon their victim and turn instead to the tasty contents of their food bowl to satisfy their hunger. It is further proof of the fact that hunting for the cat is as much a hobby as a predatory function.

Leaving a dead mammal in favour of the latest cat food brand may therefore be understandable, but as you so rightly say, finding live prey in your dining-room can be something of a shock and the sight of a cat playing with its victim can be very distressing. It is interesting to note that wild cats do not play with their victims in this way: such behaviour is the hallmark of an inefficient hunter. The rapid movement of the bird or small mammal activates the cat's inbuilt desire to chase, but the final dispatch of its victim is something that the cat must learn to perfect. To this end hunting mothers will bring half-dead prey back to the nest and encourage their kittens to practise the lethal nape bite (see chapter 5, p. 52).

These youngsters will rapidly develop the art of the quick dispatch, but for kittens whose mothers do not invest time teaching them these skills, hunting may never progress beyond the initial chase and capture. Once the prey has been caught these cats will quickly lose interest in their stationary prize and abandon it in favour of the next creature that moves rapidly across their field of vision. However, if the prey begins to struggle it will continue to fuel the desire to chase and a game will ensue. As the victim becomes weaker some cats will attempt to make it move and bat it around with their paws or even fling it into the air.

Such behaviour, which appears extremely cruel to the human observer, is a consequence of our intervention in feline reproduction, since in the wild such incompetent hunters would not survive long enough to reproduce and pass on their lack of skill to future generations. However, we supply our cats with food and protection so that the ability to kill is no longer necessary to survive, and poor hunters are able to reproduce. The result is a population of cats who never quite master the art of swift and silent murder but instead play hide-and-seek with their victims under the table.

Punishment

Whether within the context of training or treatment, punishment has no place in our interactions with domestic cats since it rarely achieves the desired effect and often does far more harm than good. It is a common misconception to think in terms of punishment as the opposite of reward, but in fact the opposite of reward is absence of reward and punishment may serve merely to confuse.

What is more, the result of punishment in cats is always unpredictable and if we take the usual human interpretation of punishment to mean physical intervention, then we shall only increase the reactivity of the cat and so increase the chance of an aggressive response. Often the animal associates the punishment with its owner rather than with the undesirable activity, and taken to extremes the result will be a cat that will do everything in its power to avoid the unpredictable humans that surround it.

Many of us will sympathise with the owner who is at the end of her tether because of a cat that is repeatedly messing in the house and I am frequently told that 'even punishing him makes no difference'. No doubt these owners have tried every reasonable approach they could think of, but pushed to the limit most humans will resort to physical punishment in the hope of reforming their errant pet. However, all punishment is completely pointless in the treatment of both indoor toileters and markers. Unless the timing is completely accurate the cat will have no idea what the punishment is about; even if, as is unlikely, the cat is caught red-handed, he will still fail to understand why he should be punished for what is after all a perfectly normal function.

For those cats whose problem stems from a case of nervous urination the use of punishment is most definitely a bad idea since it will increase the cat's level of nervousness, especially in the owner's presence, and make the condition ten times worse.

Having accepted that punishment is not the answer, we do need to recognise that certain behaviour problems in the feline world require some form of intervention to stop our beloved pets from doing something that they really should not. My own cat has been known to leap up on the work surface next to the cooker and attempt to drink chicken soup as it simmers on the ring! Not only is this behaviour undesirable from a hygiene point of view, it could also be highly dangerous for my cat, not least because he is long-haired and I cook with gas.

In such cases the important thing is to make the intervention more startling than punishing, and to ensure that it cannot be traced to the owner. In other words, our negative conditioning must be seen by the cat as a direct consequence of its own actions. The intervention should be unpleasant but never painful, and the use of water and sound can both be most effective.

In the case of my cat jumping onto the work top, I must so contrive it that his action results in some unpleasant experience, be it a shower from a well-aimed water pistol or a shock from a falling saucepan lid left balanced precariously on the surface. Either way the cat will be reluctant to go back for fear of what the work top might do next, and instead will come to its loving owner for comfort and reassurance.

Occasionally you may catch your cat in the act of performing some undesirable behaviour but have insufficient time to reach for the water pistol; in such cases a sharp hissing noise, which in feline terms is interpreted as a warning, will often be sufficient to stop it in its tracks.

R

Redirected aggression

On occasions a cat may be seen to lash out at innocent victims in what appear to be totally unprovoked attacks, and many owners who have been the recipients of such aggression find it understandably distressing. However, when the situation is analysed more carefully it becomes apparent that the cause of the aggression was not the owner after all but rather someone or something which the cat was prevented from lashing out at. Thwarted in its attempts to fight back the cat simply redirects its aggression onto the nearest available target, and often it is the owner who gets the full force of the attack.

These cats can sometimes inflict quite severe injuries on their victims, and this may be a result of the time delay between the incident that induced the aggression and the venting of the cat's anger. The delay allows the cat to fuel its aggression with excitement and so increases the intensity of the attack.

Question

My cat is normally a docile and loving individual, but when I take him to the vet his personality changes completely. While the vet is examining him I hold on tight and he stays motionless on the table, but as soon as the vet turns away, having given the necessary injections Sammy lashes out at me with claws and teeth as if it were I who had hurt him. I tell him that it was the vet and not me but he doesn't seem to understand. He has been quite poorly over recent months so I have had to take him to the vet regularly and I am worried that he is soon going to hate me altogether. How can I convince him that I am not the one with the needle?

Answer

It is always distressing when a much loved pet turns on his doting owner, but it is made much worse when the attack is unjustified. In your case your upset is compounded by the fact that you were taking Sammy to the vet because you care and want him to receive the very best attention. Understandably you are concerned that your relationship with your cat should not suffer, but it is important not to be too anthropomorphic in your perception of the situation.

Sammy is not consciously holding you responsible for the momentary pain inflicted by the needle but rather is redirecting his aggression towards the vet onto you because the vet is not available! Once the dreaded deed has been done and the injection has been administered, the vet turns away and you are the one who gently tries to reassure Sammy and give him affection. Unfortunately at that precise moment he is not in a cuddling frame of mind.

When cats are agitated in this way it is always important to handle them with caution, even if you know them extremely well. Often it is better to allow the cat to escape from the original provoking stimulus and give it time to calm down before approaching. This will mean putting Sammy straight back into his basket once his treatment is complete rather than trying to nurse him in the consulting room, but remember that he may still be agitated when you first arrive home, and putting your hand into the basket to pull him out could well result in a badly scratched hand. Instead you should allow him to leave his basket in his own time and delay your affection until Sammy is settled once more and recovered from his ordeal.

This is the short-term solution to your problem, but in the long term it would be good to remove the fear that induced the aggressive reaction in the first place. Treatment for this is described under Fear aggression, p. 120.

Rubbing

Of all the marking techniques employed by the cat, this is the one that humans find the most acceptable, and indeed we encourage it as a sign of affection from our pets. We interpret the friendly rubbing around our legs as a plea for physical contact and respond by lowering our hand and stroking in return. A desire to maintain physical contact with those close to them is part of the reason for this feline behaviour, but there is also far more to it than that (see Greeting, p. 128).

When a cat meets another familiar and friendly individual it will rub against it with its head, flank and tail and lift its tail to allow its companion to investigate its anal region. Not much of a greeting in our eyes, but in feline terms it is the equivalent of a warm shake of the hand or a peck on the cheek.

Cats possess a number of special scent glands at various locations on their body and rubbing is used to spread their secretions onto the person, cat or indeed object with which the cat is interacting. The glands on the chin, temples and corners of the mouth are used during the act of head-rubbing and the cheeks also are believed to produce scent secretions. When we stroke our cats they invariably respond by rubbing their heads along our hand, thereby marking us with their own scent and picking up some of ours in return. This helps to form close social ties between pets and their owners and establish a common scent by which to differentiate friend from foe.

This mutual rubbing is part of the complex social language of the cat and it is interesting to note that cats that are allowed access to outdoors will rub up against their owners more often than exclusively indoor cats. Single cats are also more likely to show this behaviour towards their human companions than cats which form part of a multi-cat household. In the context of feral cats, mutual rubbing, or allorubbing as it is called, is thought to be one of the main behaviours which hold the group together, but far more research is needed to elucidate the true significance of this behaviour.

Question

I have been told that when my cat rubs up against me he is depositing his scent and communicating with me in an affectionate and accepting sort of way. If this is the case, why does he rub up against my washing machine in exactly the same way whenever he returns home from one of his many hunting expeditions across the local fields? He seems to be just as pleased to see this inanimate object as he is to see me? Surely he is not being affectionate to a piece of electrical equipment?

Answer
This is a good example of how an anthropomorphic interpretation of feline behaviour can result in unnecessary feelings of rejection. How insulting to be put on a par with your washing machine!

As you so rightly state at the start of your letter, rubbing is a method of feline communication and a means of distributing scent. What your cat is doing when he rubs up against the washing machine is simply anointing his territory and those objects within it so that he can be reassured that he is safely home again. It is important for the smell-orientated feline to establish a scent profile within its territory and for everything contained within his immediate environment to be marked as belonging to it. This helps to increase the cat's confidence and feeling of security by surrounding him with his own familiar scent, and as a result he feels able to relax completely free from anxiety.

After he has rubbed up against various objects in the house, the cat will usually sit down and groom himself, thereby taking time to read the signals that he has picked up and further reassure himself that all is well. Cats do not limit such behaviour to the home, and if you watch your cat in the garden you will see that he regularly rubs up against fence posts and branches of trees or bushes that hang over his paths through the territory. By rubbing his head along these objects he smears them with his scent-filled secretions and marks the garden as being his. These scent marks not only serve to increase his confidence but will also be read by passing cats, so letting them know that this patch is occupied.

S

Scratching

In addition to its function of conditioning the claws, scratching is one of the four methods of marking used by cats for the purpose of communication and is explained in chapter 5 under 'S'. It is a natural behaviour and causes no problems until it is carried out indoors and directed against some valuable piece of furniture or expensive wall-covering.

Question

I would not consider myself to be particularly house-proud but I am beginning to feel a little upset by the damage that my cat is causing to both the furniture and the wallpaper with his constant scratching. Punishing him does not seem to have any effect and short of excluding him from the sitting-room I don't know what to do. I am reluctant to shut him out as he is very much part of the family and enjoys being cuddled in the evenings. Have you any suggestions?

Answer

The first point to consider when faced with a cat that is scratching indoors is the motivation for the behaviour. If the scratching is performed in just one or two distinct places, than it is likely that the cat is merely exercising and conditioning his claws. These cats have failed to transfer their behaviour onto outdoor surfaces such as fence posts and tree bark and instead use the corner of the sofa and the back of the chair.

Treatment is aimed at redirecting their clawing onto a more acceptable substrate and is usually quite successful. Various scratching posts are available in the pet shops, and indeed it is fairly easy to make your own, but it is not advisable to cover it with fabric or with carpet, which may inadvertently encourage the cat to scratch on similar textures elsewhere in the house.

Having decided on a suitable scratching surface, you should place it in front of the affected areas in the house, and once the cat has begun to use the new surface it can then be moved to a separate, more convenient location. If you want to encourage the cat to transfer its scratching behaviour to outdoors, it is possible to place tree bark in front of your sofa or chair and so educate your cat to use trees in future.

One of the most important things about scratching posts is their height, since a cat will need to scratch at full stretch in order to exercise the clawing apparatus adequately. Posts therefore need to be of sufficient height to allow this if you are to avoid him simply returning to the backs of the chairs or even the walls.

In some cases of indoor scratching, the motivation for the behaviour is more complex and the marking function of this visual and scent-giving signal is the root of the problem. These cats will tend to scratch throughout the house and especially around points of conflicting scents such as doorways and windows. It is important to identify the cause of the cat's feelings of insecurity, and treatment is aimed at restoring the image of home as a safe haven. These cats need to be reassured so that they no longer feel the need to surround themselves with their own scent and these cases should be treated in the same way as the indoor sprayer (see Spraying, p. 191).

Whatever the cause of the problem, punishment will never be effective and may even make things worse. If the cat is merely conditioning his claws he will not understand why you are chastising him for something perfectly natural, and if he is marking because he feels insecure, punishment will only serve to confirm his fears.

Self-mutilation

It is rare for cats to exhibit this behaviour and when it is seen it is usually in response to a very severe form of stress or one that is chronic in nature. Far more research is needed if we are to begin to understand better the mechanisms involved, and at present we are often at a loss to explain it. It may be seen as an extension of the over-grooming responsible for cases of psychogenic dermatitis (see Dermatitis, p. 106) but it need not necessarily involve grooming behaviour. There certainly seems to be some neurological component to the behaviour, and the fact that these animals fail to display any reaction to the undoubted pain they inflict on themselves suggests that their behaviour may in some way lead to the release of naturally occurring pain-killers.

Question
For some time now our Burmese cat, Murphy, has periodically clawed at his mouth. This is no ordinary pawing but an out-and-out attack on himself which results in horrific tears to his tongue. I am sure that he must be causing himself excruciating pain but he continues to mutilate himself in this revolting way. The vet suggested that there might be some problem with his mouth, but when he examined him under general anaesthetic there did not appear to be anything wrong. Could this be a psychological problem?

Answer
This is a very rare case and would be classed by behaviourists as an Obsessive Compulsive Disorder or OCD. These problems are extremely difficult to treat, simply because it is so difficult to establish a definite cause. It is likely that the cat is reacting to a severely disturbing or chronically on-going form of stress, but other possible causes may include such things as sensitivity to a component of the diet. Obviously the term stress is a wide-reaching one, and anything from seemingly insignificant changes within the household to major disruptions such as the introduction of a new cat could serve to activate this sort of behaviour.

In the short term the use of sedatives such as valium, under the close guidance of a veterinary surgeon, may be indicated, but the ultimate aim would be to identify the source of the anxiety and remove it. In reality, however, it is impossible to remove all likely environmental and managemental stressors, and drug therapy, other than sedation, may be the most helpful approach. As yet the research into the use of drugs for the treatment of these animals is by no means complete and Murphy is definitely a candidate for more detailed investigation.

Separation anxiety
This is a condition that stems from too close a dependence between animal and owner, and not surprisingly it is far more

rare in the cat, which retains such a high degree of independence, than it is in the dog. In fact it is hard to imagine that any feline would allow itself to become so inextricably linked with someone that it could not cope with being parted from them. Surely, for a creature that is fundamentally a solitary hunter, such problems should never occur.

However, we know that there are cats that become overattached to their owners and that these individuals can suffer behaviour problems when they are forced to cope alone. Such cats are likely to indulge in extremes of infantile behaviour when in the company of their owners (see Juvenile behaviour, p. 149) and may even follow them around the house constantly demanding excessive amounts of physical contact. When left alone they may hide themselves away in a dark corner or spray the house to surround themselves with familiar and reassuring scent.

Some extremely anxious individuals have been known to engage in self-mutilation when separated from the security that their owners supply and others may indulge in more bizarre behaviours such as wool eating.

However the problem shows itself, treatment is essentially the same and involves helping the cat to grow up. Often the problem arises as a result of excessive encouragement from the owners and treatment will involve non-punishing rejection of the cat's demands for attention and complete restructuring of the cat-owner relationship. This is not only difficult for the cat but also requires determination and firm resolve on the part of the owner. It is never easy to detach oneself from those that are close to you and it is no different in these situations. However, by assisting the cat to learn how to stand on its own four feet the owner is opening up a life of endless opportunities and adventures that an overattached individual would never experience. This treatment therefore aims to help the cat get maximum enjoyment out of life and is very much a case of being 'cruel to be kind'.

Spaying

By far the majority of our domestic female cats are neutered either at puberty or after having one or more litters. This is not only a reliable method of preventing large numbers of unwanted kittens but also removes hormonally controlled behaviour from the animal's repertoire. The piercing cries of the calling cat will no longer disturb the neighbourhood at night, and for the owners of particularly vocal breeds this may constitute a major advantage (see Calling, p. 95). Other than the removal of specifically sexual activities, spaying has far less marked effects on feline behaviour than castration and in general spayed cats will behave like intact females in anoestrus (that is, outside the breeding season).

Spraying

The importance of urine spraying as a marking behaviour in felines has been discussed in the 'U' section of chapter 5. It is a perfectly natural behaviour of all cats, whether male or female, neutered or entire, and the most common misconception is that it is only performed by un-neutered males. I receive countless phone calls from distraught owners who have seen their female cat spraying and presume that there must be something drastically wrong with her!

Spraying is a vital way of conveying messages to other cats in the area and is usually performed against fence posts and bushes, leaving an unmistakable mark conveniently positioned at nose height for easy reading! Most owners who have moved house will have watched their cat carefully spraying the perimeter of the new garden when it is first let out, both to surround itself with familiar scent and to let the local population know that it has arrived. In common with all marking behaviours, spraying is perfectly acceptable when carried out in the open air, but when it becomes a regular occurrence inside the house it can be a major problem. Eventually the smell of cat urine becomes noticeable throughout the house and owners steadily become less and less understanding.

The end result of this behaviour may be the depositing of cat urine in unacceptable places, but it is essential that spraying is not mistaken for indoor toileting which is a very different behaviour problem. Establishing whether you are dealing with a marking or house-training problem is the first step when faced with these cats, since both the motivation and the treatment for these behaviours will be different.

When spraying is first noticed in an adolescent male cat it is likely to be associated with the surge in male hormones and can be reliably treated by castration at this stage. Indeed, castration of entire male cats who are spraying results in an immediate end to the behaviour in eighty per cent of cases, regardless of the age of the cat.

However, when spraying becomes established in female or neutered cats it is likely to be the result of insecurity: far from being the mark of an over-confident individual, it is likely to be associated with the more withdrawn and incompetent members of the species. The presence of competition is likely to increase the incidence of spraying behaviour and in multi-cat households spraying by at least one member of the group is not uncommon. However, many people do manage to keep enormous numbers of cats under one roof without encountering a spraying problem. It may be that when the number of cats exceeds a certain threshold those incompetent individuals who would be prone to spraying do not do so for fear of drawing attention to themselves.

In cases where spraying appears to begin very suddenly, to be linked with problems using the litter tray or to be associated with any indication of pain, it is vital that the cat is taken to a veterinary surgeon as soon as possible. Cats may adopt a standing posture to pass urine when they are suffering from cystitis or a condition called Feline Urological Syndrome (FUS). This complaint results in the partial or sometimes total blockage of the urinary tract, preventing passage of urine and causing considerable pain to the sufferer. Early treatment is needed and this must be followed by careful supervision and manipulation of the diet in order to prevent recurrence of the problem. These cases are a firm reminder of the need for veterinary involvement in behaviour problems and close co-operation between veterinary surgeons and pet behaviour counsellors.

Question

After saving up for some considerable time I have recently had my kitchen completely refitted and I am very pleased with it, but it has acquired a distinctly feline odour over recent weeks. My cat was quite upset by all the comings and goings when the workmen were around, but generally she just stayed out of the way. I didn't think any more about it until I began to notice marks on the kitchen cupboard doors and a smell of cat urine in the air. I have never actually seen her spraying but that is obviously what she is doing. I love my cat very much but I also love my new kitchen and ideally I would like to be able to keep them both!

Answer
Thankfully, for most of our cats the home is a secure base where they feel calm and protected and therefore they see no need to identify it further as theirs. These cats view their owners as their maternal protectors and rest happily in their company, sure in the knowledge that they are safe. Spraying has no place in the home as far as they are concerned and they limit this activity to the garden and territory boundaries.

This is the normal situation, but for some cats, such as yours, something happens to call the security of the home into question and cast doubt on its ability to protect them from harm. It is likely that your cat has always spent quite a lot of her time in the kitchen, and in common with most cats she was probably fed there and may even have had a bed in the corner. Then in come some strange men, rip out everything that she sees as hers and replace it with new cupboards and equipment which carry a vast supply of unfamiliar and challenging scents. Looking at it from the cat's point of view it is hardly surprising that she feels somewhat insecure and in need of reassurance.

The fact that she does not spray in front of you is an indication that you act as a source of security, and while she is relying on you to fulfil your protecting role she feels confident enough not to spray. In this case it is not difficult to identify the cause of the problem and the cat makes it even easier by directing her spraying against the source of her anxiety. Such specific targeting is common, and many cats will spray on plastic carrier bags or new pieces of furniture because of the unrecognisable scents they bring into the house, which disrupt the cats' olfactory senses.

In cases where the security of the home is weakened by the arrival of a new baby, or perhaps even a new husband (see Attention-seeking, p. 84), the spraying behaviour may be specifically directed against their clothing and belongings. Not all cases of spraying have such a clear-cut cause, and for some it is a combination of stressful influences which pushes the cat over its emotional threshold. These cases are harder to treat, simply because it is impossible to remove the challenge or indeed help the cat to come to terms with it.

Treatment of the indoor sprayer involves increasing the cat's perception of its home as a secure den as well a breaking the spraying habit. The fact that the spraying usually occurs in the owner's absence means that punishment has no place in the treatment programme, and even if the cat were caught in the act, punishment would be counter-productive since it would increase anxiety and make spraying more likely.

In cases where the spraying results from structural changes to the home, it is sensible to deny the cat any access to those areas unless it is supervised by the owner, thereby giving time for the alterations to become accepted. Reducing the size of the territory by employing an indoor pen can help to increase the cat's confidence. The pen should contain a warm, covered bed as well as a litter tray and food supply. From the protection of the pen the cat can come to terms with its environment and gradually be introduced to more and more rooms in the house, culminating in the kitchen where the specific challenge is located.

The first ventures into any room without the pen should be supervised, so that the owner can act as a security bridge and the cat can steadily increase its perception of the home as a safe resting place. Cat flaps may be extremely useful inventions from a human point of view, but for a cat that feels under threat this hole in the outer defences of his home can add to his apprehension and blocking the flap up at least temporarily may be beneficial.

Having removed the cause of the spraying behaviour or enabled the cat to come to terms with it, the next step is to break the habit of urine spraying. To achieve this, it is of paramount importance that all areas that have been anointed should be thoroughly cleaned. Scent marks are known to convey vital information to the cat, and the smell of an old and decaying scent mark will induce the perpetrator to top up the messages with an extra dose of urine, even if the original problem has been overcome.

For most owners, cleaning is all about removing the smell and replacing it with a clean and hygienic odour. Consequently they use one of the wide variety of household cleaners which leaves the home smelling of ammonia or chlorine. Unfortunately both of these compounds are constituents of cat urine, and although these products may be indicative of cleanliness and sterility to the human nose, to the cat they indicate the presence of a rival marker. As a result the cat will be eager to counteract the intruder's scent with his own and will deliberately spray once again.

In the same way, strongly perfumed detergents or pot pourri air fresheners may serve to confuse the scent-orientated feline: these smells may be perceived as a challenge and so lead to an increase in spraying activity. This game can go on *ad infinitum*, with the owner getting more and more frustrated as attempts at cleaning up the house are met with increasing bouts of marking.

The answer is to use a cleaning regime that will remove the smell of cat urine but will not substitute it with a challenging odour. The most successful method involves firstly cleaning the area with a warm solution of a biological washing powder in order to remove the protein components of the urine, and then rinsing with cold water and allowing the area to dry. This should be followed up by scrubbing with an alcohol such as surgical spirit, in order to deal with the fatty deposits, and after cleaning, the area must be allowed to dry thoroughly before the cat is given access to the room again.

The two-fold approach is necessary since it is thought to be the decaying fat deposits in particular which encourage cats to freshen up their ageing scent marks. It is possible that this regime may result in the removal of dye from certain fabrics and it is important to do a test patch before treating whole carpets or sofas in this way.

Although this cleaning regime has been very successful in the treatment of spraying cats, there may be times when the odour is so ingrained in carpets and other furnishings that the only hope of successful treatment is to get rid of these articles, which can of course involve the owners in quite considerable expense.

Once the area has been cleaned the cat will not feel it necessary to over-spray, but if a new challenge were to occur he might be driven to spray again and it is important to deter him. Certain positions in the house are more likely to be selected as scenting posts and these include points of conflict such as doors and windows. Countless preparations which are designed to act as deterrents to the spraying or urinating cat are on sale in pet shops throughout the country, and in addition many people will be able to tell of their own original and often ingenious methods of preventing this behaviour.

A simple deterrent, and one that is most effective in the majority of cases, is food. Cats will rarely spray in the vicinity of their meal and it may be that the presence of food also acts as a reassuring signal and helps to increase the cat's confidence. From a hygiene point of view, dry food is more suitable for this purpose than the canned variety, and it may be helpful to stick the food to the bottom of the dish in order to prevent the cat from eating it and then spraying!

Hopefully, by applying these treatment measures your cat will be able to come to terms with your new kitchen and in time even appreciate your choice of design.

T and U

Territorial

We all recognise that the cat is a territorial animal and in chapter 5 we looked at the basic instincts that determine their territorial behaviour. Unlike the feral cat who selects his home base and maps out his own territory, the domestic cat has to establish himself in a territory that is thrust upon him. Often this will involve vying for position with already well-established residents, and most owners can expect their cats to receive some degree of injury during the first few weeks in a new home, even if it is only a torn ear margin or a scratched nose.

Once again the cat's natural tendency to avoid confrontation comes into play, and rather than wade in with claws flying, most will quietly and slowly integrate into the existing population. Over-emphasising the cat's territorial tendencies can lead people to think of it as an asocial animal, but numerous studies carried out by behaviourists on feral cat colonies prove that a highly complex social scene can develop.

In general the size of home ranges and the density of feline inhabitants will be directly affected by the level of food availability. Feral cats that live in urban areas are provided with rich pickings of discarded food from dustbins, as well as tasty morsels from the kindly souls who take pity on them and leave out bowls of cat food in their back gardens. Consequently the home range can support a far greater number of cats than a truly rural range where food consists of live prey that must he hunted and killed before feeding can take place.

Similarly, cats that set up home close to farm buildings or near to refuse dumps will find that the food supply is constant, if slightly less inviting than in suburbia, and the density of cats found in these locations will be somewhere between the other two.

When we come to the situation of the domestic cat, who is provided with an abundance of palatable and easily digestible cat food on demand, it is not surprising that the home range requirements can be correspondingly small, and we further assist this process by neutering the majority of our domestic cats and so removing any sexual determination of territory sizes. As a result, housing estates can quite safely accommodate large numbers of cats living relatively peacefully side by side and this is just as well in view of the increasing popularity of the cat as a domestic pet and the tendency for more and more multi-cat households in modern day suburbia.

If this were the whole story cat fights would be a rare event, but those of us who live in areas of high feline populations know that this is unfortunately not the case. It only takes one particularly aggressive cat or an unneutered male to move into the area, and the peace of a stable cat population can be suddenly shattered. Many is the time when clients have complained to me about the arrival of a despot into their quiet cul-de-sac as they bring their cat to the surgery to have another abscess lanced and yet another course of antibiotics. In some cases this newcomer is an unneutered male who is simply responding to his hormonal drive, and provided that the owner can be persuaded to have him castrated the problem will usually resolve itself (see Castration, p. 98).

However hooligans exist in the feline world as much as they do in our own, and territorial aggression can become a real problem for the owners of such characters. Treatment must involve full co-operation between local residents and the methods used have been described in detail under Fighting (p. 123).

Whereas territorial aggression in canine terms conjures up images of dogs attacking postmen and paper boys, any mention of feline territorial aggression will immediately bring to mind outdoor skirmishes across the garden fence. However, for cats, too, their home base is an important part of their territory and one that should be protected. In normal circumstances cats do not have to cope with invasion of this core territory by other felines unless a local rival begins to break in via the cat flap, but occasionally owners try to introduce new residents and come up against fierce opposition. This

Occasionally owners try to introduce new residents and come up against fierce opposition.

form of territorial aggression has already been dealt with under Introductions (p. 145).

Territorial aggression directed towards humans is far more rare in feline terms than it is for their canine counterparts, but nevertheless some owners do indeed find themselves threatened by their cat when they attempt to move him from his favourite chair or disturb him by brushing past him in the hall. In some cases these cats may be displaying dominant behaviour which would be more readily attributed to a dog and treatment must be aimed at restructuring the relationship between cat and owner and decreasing the animal's perceived rank within the family unit. Ensuring that the cat has access to outdoors and is able to express his full range of natural behaviour patterns may help by decreasing the atmosphere of tension and enabling him to be more amenable to treatment.

Toileting

See House-training, p. 137 and Indoor toileting, p. 141.

Training

For most people the concept of training is one that belongs to the canine world, but the presence of cats in films and television advertisements is proof that it is possible to train our cats if we so wish. Cats are well able to learn, and do so remarkably quickly provided that basic training principles are applied.

First, it is important to study natural feline behaviour so that we can more easily predict reactions and exploit their natural responses within our man-made context, and secondly we must remember the rule of reward rather than punishment. In the past we have repeatedly failed to apply these rules with our dogs and have still managed to train them in spite of ourselves, but if we try to take the same ham-fisted approach to training our cats we are sure to fail. Training needs to be positive and the trainer needs to have an endless supply of patience and understanding if he is to succeed. Quite simply, the cat should never be allowed to fail, and when it does perform the required behaviour it must be rewarded appropriately, quickly and consistently.

Those who train cats for work with both television and films will tell you that it is impossible to force a cat to comply with human demands but that it is possible to train them provided the training programme is based on an interactive relationship and a kind and patient approach. It is helpful to start training while the cat is still very young and is willing to respond to new experiences and situations, and it will be an added advantage if the kitten is naturally curious and highly responsive.

The reward used in training must be perceived as such by the cat, and it can prove far more difficult to find an appropriate reward system in feline terms than for the more readily trained dog. Dogs will often respond well to food irrespective of whether they are actually hungry, but most cats are less impressed by tasty morsels and tempting titbits. Equally, praise and affection do not hold the same appeal for an essentially independent creature who can take or leave our attention at the best of times.

It is therefore important to adjust the reward to meet the individual requirements not only of the cat but also of the particular situation, and flexibility is essential. Timing of the reward is the other crucial factor and delays of just a matter of seconds can lead to unnecessary failure. The reward needs to be associated directly with the action and must therefore be delivered immediately, and consistency of reward will help to ensure that the cat learns quickly. By applying these principles it is possible to train cats to perform a variety of tricks, but for the average cat owner there is little benefit in that. However, training can be relevant to the ordinary domestic cat, for example in facilitating their acceptance of that popular invention, the cat flap.

Question
Until relatively recently I lived in a second floor flat and my cat was able to gain access through an open window, although quite how she managed it was always something of a mystery. Anyway since then I have bought a house and for security reasons I do not think it would be sensible to leave a window open while I am out at work all day. However, Toffee is used to her freedom and I cannot bear the idea of her being shut in all day; equally I do not like to think of her being shut out in the cold and the rain. As a result I went to the local pet shop and invested in one of these super cat flaps complete with electronic tag for Toffee, so that she can use the flap but not allow other cats into her home. I really thought that I was being generous but Toffee is less than impressed by my gift and flatly refuses to use it. Is there anything you can suggest, since I feel that she is unnecessarily missing out on life at the moment?

Answer
Life would be so much easier if cats could read, but unfortunately they can't and Toffee must be forgiven for not realising that the cat flap is a very well-intentioned and generous gift from her loving owner. One thing is certain: Toffee is not alone in her reluctance to use a flap, and indeed many owners have written to me with this frustrating problem. You do not say how old Toffee is, but judging from your other comments I would assume that she is not particularly young, and it is always more difficult to introduce novel experiences in adulthood than when kittens are small and eager to learn. However, provided that you have a good supply of patience it is very likely that Toffee will learn to use the cat flap, and once she does she will no doubt thank you for your thoughtfulness.

You need to remember the principles of training and base your approach on the use of rewards and abandonment of punishment. Forcing Toffee to use the flap by physically pushing her through it is likely to be counter-productive and may even make her so frightened that she is reluctant to approach the back door in future. First you should prop the cap flap open and put up with the draughts for the time being! With the flap wide open Toffee can be gently coaxed through the hole by calling her reassuringly and offering food so that she comes to accept the fact that it is possible to enter and leave her core territory at this point.

Once this first stage has been successfully completed the flap should be gradually lowered so that Toffee can see daylight the other side and only has to apply gentle pressure to open the flap sufficiently to get out. If at any stage she refuses to attempt to use the flap, you should retrace your steps and leave it propped open for a little longer.

Eventually you should reach the point where Toffee learns to push the cat flap open from a closed position, but bear in mind that some of the more expensive flaps fit extremely tightly when shut and need a fair amount of force from the cat in order to open them. Take your time and give Toffee plenty of encouragement, and I am sure that your investment in the cat flap will not have been wasted.

Urination

See Indoor toileting, p. 141.

V

Veterinary surgeons

Pet behaviour therapy is a rapidly expanding area of interest in this country, and as a practising veterinary surgeon I am keen to see it incorporated into the vast array of services that our profession provides for its patients. Every veterinary surgeon admitted into the Royal College of Veterinary Surgeons makes a vow that 'my best endeavour will be to ensure the welfare of animals committed to my care'. There can be no doubt that behaviour therapy is as relevant to animal welfare as any of the already well-established medical or surgical disciplines, and I fervently believe that animal health must be seen to incorporate both physical and mental components.

Just as with human health, the symptoms of physical and psychological ailments in animals are not mutually exclusive and it is essential that full medical examinations are carried out first to ascertain whether any change in behaviour patterns is the result of some underlying medical complaint.

With the already huge diversity of disciplines in the veterinary profession it is obviously not possible for every vet in the country to have a specific interest in behaviour, nor is it feasible for every practice to offer behavioural consultations at their premises but there does need to be an increasing awareness of just how relevant behaviour is to the whole concept of animal health. Thankfully such an awareness is undoubtedly growing within the profession and veterinary education at undergraduate level is beginning to incorporate this field into the already bursting curriculum.

As a result, many veterinary surgeons are able to provide their clients with both medical treatment and behavioural advice, but in the context of a busy practice it may be impossible to spend sufficient time with individual clients in order to establish the causes of problems and institute suitable treatment programmes.

When dealing with cases of behaviour problems we are often dealing with what is fundamentally a breakdown in a relationship, and that can carry with it feelings of inadequacy and failure. In many cases the problems run deep, and tensions within the family may be having marked effects on the cat's behaviour. In these cases it is essential that the problem is dealt with compassionately and sensitively, and counselling is very much part of the job. Time-pressured veterinary surgeons must be able to refer their clients to professional people in the sure knowledge that they will receive not only good, sound advice but also the necessary counselling. The Association of Pet Behaviour Counsellors, which is a nationwide and indeed international network of experienced professional behaviourists available exclusively on referral from veterinary surgeons, provides such a service.

I am proud that the veterinary profession is held in such high regard by the public, and honoured that they trust us with the well-being of their beloved pets. In return for that trust we owe it to owners and pets alike to take behaviour problems seriously. As veterinary surgeons and behaviourists work closely together, we can all aim for that ideal pet-owner relationship, to the mutual benefit of man and animal.

Vomiting

See Emesis, p. 112.

W

Wandering

All of us who choose to share our lives with companions of the feline variety have to accept that, unless we decide to make them exclusively indoor pets, we shall never have control over them in the same way as we have over our dogs. We may attempt to decrease their desire to wander away from the nest by neutering them at an early age, but when the call of the wild, or indeed the call of a more comfortable pad down the road, takes hold, we feel powerless to make them stay. Although we may accept the independence of our pets and indeed admire them for it, there are times when the fact that they use our houses more like a hotel than a home begins to cause us some distress.

Question

I am becoming increasingly concerned by the fact that my cat is spending more and more time away from home and becoming steadily less and less interested in me when he does eventually come indoors. When I first got Otto he was always there to greet me when I arrived home, and after eating his dinner he would happily curl up on my knee and sleep the evening away. Then, about six months ago, I was temporarily transferred in my job and had to travel forty miles each day to get to and from the office. Obviously I had to leave the house very early in the mornings and it was quite late before I got home at night, and sometimes I didn't set eyes on Otto for days at a time.

I knew that he was about by the cat hairs on the chairs and the muddy paw prints on the kitchen floor, and sometimes in the middle of the night he would creep upstairs and lie on the foot of my bed, but I could count on the fingers of one hand the number of times we sat down together to watch TV in the evenings.

I accepted that there was very little I could do about the situation and comforted myself with the fact that the arrangement was only temporary. However, it is now three months since I have been back at my own office and the situation with Otto is not improving. He rarely comes home before 11 o'clock at night, and when he does it's only to demand a feed before turning round and going back out again. When he is indoors he is reluctant for me to pick him up and I have to physically restrain him if I want him to sit on my lap. I am finding this deterioration in our relationship extremely distressing and would like to know how I can win back my pet's affection.

Answer

Your story is a particularly sad one since your absence from the house in the evenings was not from choice and in many ways, if you had chosen to go out every evening to wild parties, Otto's rejection would be easier to take. Unfortunately he is unable to understand your motive for leaving him alone so much and as far as he was concerned your house simply failed to provide the love, warmth and affection he had come to expect. With little reason to return to his core territory Otto began to spend more time out and about and may even have found another house where people were constantly available and willing to indulge him.

The fact that he continues to return home at night and demand food from you is a good sign that he still feels tied to his territory and we must aim to build on this and increase his perception of home as a safe and secure base rather than a convenient stop-over. If you are able to discover where Otto is going and who, if anyone, is feeding him, it will certainly help if you can persuade them to refrain from encouraging him and even ask them to actively discourage him in order to help with your efforts to win back his affection.

To start with you should invest in some of Otto's favourite food and set specific times when he will be fed. This will help to re-establish home as a secure and predictable place, and by ensuring that these times coincide with your availability you can make certain that when he comes home to eat you are on hand to give him plenty of love and attention. In this way you can work to re-establish your role as his provider and protector and gradually deepen the bond between you.

Do not worry too much if there are occasions when you cannot be home at the required time, since you can invest in one of the timed automatic feeders so that Otto can help himself to his dinner and still maintain a predictable routine.

If Otto is reluctant to allow you to pick him up do not force the issue since this will only make him feel threatened and so be counter-productive. If it is possible for you to take some leave it may well help for you to be around the house more and you could always shut the cat flap and keep Otto confined to the house so that he gets used to being in your company. I would not advise you to keep it shut permanently, however, since Otto is obviously used to his freedom and is unlikely to react well to prolonged confinement.

As you take time to be available to Otto and greet him enthusiastically when he returns home, it will hopefully not be long before he begins to respond and spends less time exploring and more time at home.

Wool eating

This is arguably one of the most bizarre behaviour problems in the feline world and one that, as yet, is not fully understood. It was first reported in the 1950s and at that time was believed to be restricted to certain Siamese strains, but work carried out by cat behaviourist Peter Neville, in conjunction with John Bradshaw of Southampton University, has shown that the problem is in fact more widespread.

Their survey revealed that this behaviour occurs in Siamese and Burmese cats as well as being seen occasionally in some other Oriental breeds. Crossbred cats were also found to exhibit the behaviour, albeit rarely, and although some of the cats in the survey who were of mixed parentage did have some Oriental genetic influences, others were apparently moggies. Females and males are equally likely to show this peculiar eating behaviour, and neutering has not been shown to have any reliable effect on its incidence. The majority of cats surveyed began their fabric eating during adolescence, and the typical age of onset was between two and eight months of age.

One of the most interesting findings is that although the majority of affected cats do start by consuming wool, few remain so restricted in their appetite. Many expand their tastes to encompass all fabrics, including cotton and synthetic fibres, and therefore fabric eating is in fact a more accurate term to describe this fascinating behaviour. Some of these cats chew and eat fabric on a daily basis, but for others the behaviour has a far more sporadic incidence.

The most sought-after delicacies appear to be various items of the owners' clothing, from woollen jumpers to underwear, and those that have been recently worn seem to hold an added attraction. In most cases the damage is not restricted to clothing, however, and towels, tea towels and even furniture coverings have all been reported as part of the daily diet for some individuals.

In many cases the amount of fabric that is consumed almost defies belief and it is amazing that with the majority of cats this material passes through the digestive system without causing any apparent harm. Occasionally, however, the material does become impacted, either in the stomach or in the intestines, creating a total or partial obstruction, and then surgery is needed to remove it. Unfortunately for some, the resulting damage to the digestive tract may be so severe that euthanasia is the only answer, but many more will continue to eat wool and other fabrics throughout their long and very healthy lives.

The financial cost of the damage caused by some of the cats is phenomenal and for some owners it becomes too much to bear. These cats may regrettably have to be euthanased but an amazing number of other owners tolerate the behaviour with remarkable patience and are not even put off the idea of owning other individuals of the same breed in future.

Question

I have owned Siamese cats for quite some time now, but my present companion is proving to be something of a problem. She has always been a very affectionate cat and has kneaded constantly while I cuddle her. She also had a habit of sucking on my jumper when she was very relaxed and I didn't really think anything was wrong with such behaviour. While I can tolerate sucking and kneading, she has now started actually to eat my woollen jumpers and create often quite large holes in them and I am beginning to find that her behaviour is rather costly to bear.

I am not the tidiest of people and have been known to leave clothes lying on the floor, but now I am having to make sure that anything of value, sentimental or otherwise, is put away before Smidgen gets her teeth into it. I feed her on one of the commercial cat foods and have always thought that they should be sufficient to meet all nutritional requirements, so why does she find it necessary to supplement her diet in this way?

Answer
The precise cause of this fascinating eating behaviour has not been identified but it would appear that it is not a result of any nutritional deprivation, so you need not blame yourself for Smidgen's problem. A variety of potential causes has been investigated and it has been suggested that neuronal disturbances involving the involuntary or autonomic nervous system may be operating to affect the control of the digestive system and induce unusual appetite stimulation; the actual details of how this may occur are by no means clear, however.

The behaviour also seems to be inherited, at least to some extent, but whilst there may indeed be a genetic component the actual trigger for the behaviour is most likely to be environmental. It may be that some form of stress induces certain individuals to consume fabric, and in the survey by Neville and Bradshaw a significant number of the cats began to show this behaviour within one month of entering their new homes. Transfer to a new household is thought to be one of the most common environmental triggers for fabric eating but others may include introduction of another cat, unexpected separation from the owner or a period of physical illness.

Some individuals appear to grow out of the behaviour by the time they are two years old and it may be that these cats have gradually learned to cope with the stress that acted as the original trigger. However, others continue to eat their way through any fabric that happens to be left lying around and often do so quite openly.

Once a cat has begun to eat a tasty piece of cardigan or other inviting garment it will often enter what appears to be a trance-like state and no amount of shouting by the irate owner will make him stop. Even in cases where using a water pistol has been effective in interrupting the behaviour the cat may simply return to the 'meal' immediately afterwards or take the article away to a quieter spot and carry on where he left off.

One possible explanation which has been offered for this bizarre behaviour is that wool eating and sucking are extensions of juvenile behaviour exhibited by individuals who fail to reach full maturity. Certainly a proportion of fabric-eating cats, including Smidgen, display a high degree of infantile behaviour towards their owners in the form of kneading, sucking and dribbling when cuddled, and some are particularly dependent and tend to follow their owners around the house.

Treatment for these cats must involve encouraging them to become more independent and adult in their approach to life, and so the periods of interaction with the owner should be decreased and shorter doses of affection should be made available only at the owner's instigation rather than the cat's.

The fact that fabric-eating appears to be more prevalent in indoor cats suggests that increasing the level of stimulation and activity would be beneficial, and ideally these cats should be allowed access to the great outdoors in order to reduce the importance of their owners and increase the possibility of encountering and investigating novel stimuli.

Preventing access to any unsupervised items of clothing or other edible fabrics may not always be easy, but it is most certainly helpful in the treatment of these individuals, and in some cases being denied their material diet for a period of a few weeks can even result in a cure. Unfortunately, though, treatment is not that simple in every case and other tactics need to be employed. Attempts to ambush the cat in the middle of its main course, using a water pistol or other startling device, may help in some cases but there is always the danger that such methods may simply encourage it to eat wool in privacy in future and make the problem even more difficult to deal with.

A more effective method is to make the fabric itself the negative stimulus by applying taste deterrents to pieces of material that are deliberately left for the cat to find. In general the traditional unpleasant-tasting products such as pepper, curry and mustard have little effect in these situations and it is better to use aromatic compounds such as oil of eucalyptus or menthol. Provided that a suitable taste is used, such techniques can have a dramatically reforming effect and put the cat off ever attempting to chew fabric again.

Despite the fact that wool eating does not seem to have any nutritional basis and the cats that indulge in this behaviour appear to have perfectly normal appetites and a healthy intake of their ordinary food, dietary manipulation still seems to be useful as part of the treatment. Increasing the fibre content of the diet by providing bowls of dry cat food in addition to the usual meal may help to encourage intake of an alternative food source in preference to fabric, and some cats can be cured by simply allowing access to an unrestricted supply of dry food. This added fibre may simply bulk out the diet sufficiently to ensure a constantly comfortable full feeling, so that the cat no longer feels the need to supplement its intake with fabric.

If this is the case, then the same result can be achieved by adding bran to the usual diet, but most cats will only accept this up to a certain point and a more acceptable alternative from a feline point of view may be to add small fragments of finely chopped fabric to the feed bowl. Indeed, some owners have opted to give their cats a supply of cheap fabric which is made available at meal times and found that the cat will take alternate mouthfuls of fabric and cat food.

These last two options may seem to be giving in to the behaviour, but if they are successful in stopping the cat from indiscriminately devouring expensive clothes and furnishing they do represent a viable approach to treatment.

One other treatment strategy which may be of use is the provision of tough chunks of meat and large meat-covered bones. The basis for this is the observation that fabric eating appears to be an expression of a prey-catching behaviour, and an outlet for the energies which would naturally be directed towards preparing live prey for consumption. The fact that we feed our cats on easily digestible and ready-to-eat foods means that they are no longer required to exercise their teeth and jaws as nature intended, and it is interesting to note that a significant percentage of fabric eating cats have little or no access to outdoors and are therefore denied the opportunity to express their natural hunting behaviour. Forcing them to spend more time chewing at gristly meat and tearing meat from the bone increases the overall amount of time spent in normal feeding behaviour and can reduce or even eliminate the desire to consume fabric as a dietary supplement.

X

In this part of the book I have tried to cover the most common questions that cat owners ask about feline behaviour, and to help at least in some small way to deepen our understanding of these wonderful creatures. As I come to the end of this problem alphabet I cannot think of any problems that fall into the category of Y or Z, so for the purposes of this book X has become the final letter.

Xenophobia—fear of strange people

There is a vast number of cats in homes throughout the country who find the presence of strangers within their territory too much to handle. These are the cats who, on hearing the doorbell, fly out of the cat flap and down the garden at a rate of knots, or hide quivering beneath the coffee table hoping that they won't be seen. For some, their inability to cope with visitors has been present since kittenhood (see Nervousness, p. 165) but for others the fear has resulted from a traumatic experience later on in life. Whatever the cause, no one would dispute that life for these cats is less than ideal and many owners become most distressed by seeing their pets in such a state of anxiety.

Question

Whisky is a six-year-old neutered male and for 90 per cent of the time he is a friendly and confident individual, but when our doorbell rings he changes into a quivering wreck and runs under the coffee table in our dining-room, refusing to emerge until the visitors have left. I cannot think of anything that could have brought on this unjustified fear and I hate to see him so obviously distressed. What can I do?

Answer
Your description of Whisky as a normally confident character would suggest that his fear is unlikely to be the result of lack of early experiences and a more plausible explanation is that he has, at some time in the past, been frightened by a visitor to the house and now perceives all strangers as people to be avoided.

The cat is a naturally cautious animal and will avoid conflict if at all possible by employing its highly developed flight response. This is what Whisky is doing when he runs for cover under the coffee table, and although your visitors would simply like to say hello and be friendly towards him Whisky is not taking any chances. Therefore he doesn't stick around long enough to find out that the threat is purely an imagined one.

Not only is his own quality of life being unnecessarily diminished, but before long you too may well start to dread that knock at the door because of the obvious distress that it causes. As time goes on and more visitors arrive at the front door Whisky is likely to spend increasing amounts of time in hiding and so decrease his perception of his home as a secure base, and indeed some owners of particularly xenophobic cats have been known to stop having visitors for fear of upsetting the cat. This would seem to be a drastic solution to the problem and it is far better for both the owner and the cat to work towards improving his competence rather than avoiding the situation altogether.

It is important in treating any form of nervous behaviour not to overwhelm the cat or make it feel cornered, and forcing a xenophobic cat to be held by an enthusiastic visitor will only serve to heighten his anxiety and increase the chances of someone receiving a nasty injury! Cats are well equipped with teeth and claws and if pushed to the limit they will use them to remarkable effect. Instead treatment should be calm and controlled and the first stage involves forestalling Whisky's attempts to avoid visitors, thereby giving him the opportunity to learn how to cope with the situation. This is best achieved by employing a cat carrying basket of the sort used to transport cats to the cattery.

The basket, with Whisky inside, should be placed in the sitting-room before visitors arrive and the first 'guests' who enter the room should be members of the family who have rung the doorbell rather than using their keys. As the doorbell sounds Whisky will try to escape, but his attempts will be thwarted by the confines of the cat basket and he will be made to face the intruder. When the door opens, however, the cat will recognise a family member and quickly calm himself, so learning that not all rings at the doorbell are necessarily bad news. By repeating this process many times, Whisky will come to associate the ringing of the bell with the arrival of a totally non-threatening 'visitor' and then well-known friends can start to enter the room accompanied by a member of the family.

These first true visitors should do no more than sit in the same room as Whisky and should not attempt to approach him or even acknowledge his presence. The idea is that Whisky needs to be given the opportunity to accept the presence of strangers in very gradual stages, and while protected by his basket, he can start to come to terms with the fact that not all people are threatening as he had come to believe.

If Whisky is particularly distressed during attempts to follow this treatment plan, you can help him by using a low dose prescription of a suitable sedative or a homoeopathic remedy which is designed to dampen his reactions and allow him to tolerate such exposure. As with any behaviour problem, it is essential that progress does not become drug-dependent, and the dose should be gradually decreased so that his tolerance of visitors can become increasingly learned.

With repeated and frequent exposure, under these conditions, to a wide variety of guests, Whisky should come to accept that strangers can enter and occupy his territory without posing a threat, and once this stage has been reached he can be helped to tolerate their presence at ever closer proximities. Guests should be asked to sit gradually closer and closer to the basket, but they should not attempt to make physical contact until he is totally confident about their presence.

The important point to remember at this stage is that treatment must only progress as fast as the cat himself can tolerate, and a great deal of patience may be needed. Once Whisky is happy for visitors to sit beside his basket he can be encouraged to accept some interaction by them and this is best achieved by using food as a mediator. It will help if Whisky is particularly hungry, since he will then be more willing to accept food from a stranger, and so he should be starved for twelve hours before the next visitor calls. Sitting beside the basket the guest should calmly and slowly offer a small titbit through the bars, while talking gently and reassuringly. Hopefully Whisky's desire to eat will be strong enough to overcome his fear, and eventually he should be able to be fed short, frequent meals by as many visitors as possible, so that he comes to view all strangers as potential meal tickets.

Finally Whisky needs to learn how to cope with strangers without the security of his basket, and for this stage it may help to restrain him on a collar or harness with an extending lead. When the guests enter the room they should offer him food and then, with Whisky held firmly but gently in your arms, you should slowly approach a particularly well-known individual. If at any time Whisky begins to show apprehension or anxiety you should stop and allow him the opportunity to calm himself. Once the visitor is next to Whisky, he or she should gently start to stroke his coat while you continue to pet him, and gradually you should withdraw your hand until it is only the guest who is doing the stroking.

It is important to pay attention to the way the cat is approached during treatment: in general, advances should always be made at the same level as the cat—that is, face to face—rather than from above, since this will seem far less threatening. However, it is important to avoid any staring eye contact, as this could be mistaken for a challenging gesture, and any loud noises or sudden and unexpected movements should also be avoided.

Do not expect to reach the stage where visitors can pick Whisky up for some considerable time, if ever. It is one thing to allow gentle physical contact but it is quite another to allow yourself to be restricted by being held, and many cats are reluctant to permit such behaviour by anyone but their close family. Provided you approach treatment calmly and slowly Whisky should make good progress, and with regular exposure to visitors in the future he will be able to live life to the full once more. However, it is sensible to remember that particularly loud or vivacious visitors may still be somewhat alarming, and if you intend to have wild parties it may be kinder to put Whisky into the cattery overnight rather than subject him to any raucous behaviour.

APPENDIX

ASSOCIATION OF PET BEHAVIOUR COUNSELLORS
At various points in this book I have made reference to this Association which was founded in 1989 and now comprises fourteen member practices throughout the United Kingdom. In 1990 I became the first veterinary member of the APBC and as its veterinary officer I am keen to establish links between the veterinary profession and behaviourists in this country. All APBC members accept cases exclusively on referral from veterinary surgeons, and if after reading this book you feel that you require more detailed and personal advice you should contact your veterinary surgeon with a view to being referred to your nearest ABPC clinic.

A full list of APBC clinics is available on request and details of the numerous publications by APBC members can also be obtained by writing to: The Hon Secretary, Association of Pet Behaviour Counsellors, 257 Royal College Street, London NW1.

EPILOGUE

During the writing of this book my cat Truffles was tragically killed in a road traffic accident. She was a wonderful companion and had been with me through many important changes in my life. A farm cat by birth, she was always a most proficient hunter and no doubt the local population of birds and small mammals does not share in my sense of loss. Throughout her life she was constantly teaching me about feline behaviour and I thank her for helping me to deepen my appreciation of such a fascinating species.

INDEX

Numbers in bold type are main headings